Advances in
Microbiology

Advances in Microbiology

Arunima Mukherjee

Oxford Book Company
Jaipur • New Delhi

Published By

Oxford Book Company

Regd. Off. "Bony Residency", Gate No. 2
Opp. Tilak Public School, Vishweswariya Nagar,
Gopalpura Road, Jaipur - 302018, India
Ph: 0141 - 2761280, Telefax: 2761381
email - oxfordbookjpr@gmail.com
website - oxfordbookcompany.com

Delhi Office: 102, 1st Floor, Satyam House,
4327/3, Ansari Road, Darya Ganj,
New Delhi - 110002, India
Ph: 011-45652440

Distributed By

KEYA INTERNATIONAL PRESS LLP
E-mail: keyainternationalpressllp@gmail.com

Published in India, 2024

ISBN : 9788189473440

Price: ₹ 1950.00

Typeset by Shivangi Computers, Jaipur

Printed and bound in India by Thomson Press Ltd.,
New Delhi

Preface

Microbiology is the scientific field that concerns itself with the study of organisms that cannot be seen by the naked eye, which include bacteria, viruses, certain algae, fungi and protozoans. The existence of microorganisms was first demonstrated in the 1660s, when Robert Hooke of England built compound microscopes and Acton van Leeuwenhoek of Holland constructed powerful lenses, which these men used to study and illustrate various microbes. The field of microbiology was rounded out in the microscope, which enabled the observation of even the smallest microorganisms, viruses, and further exploration of other microbes. As proceeding researches have proven, microbiology has had immense applications, especially in medicine and the food industry.

This book, one in the series of books on various sciences and their contemporary status, concerns itself with the recent advances in microbiology. Besides, its use in genetic research, modern microbiology has contributed significantly to the study of disease and to research in agriculture and environmental health. The book, however, is not a mere inventory of the contributions of microbiology to contemporary science, but also a manual which deals with the discussion of the principles, practices and prospects of microbiology today. In addition, it also incorporates within its pages a discussion and incisive book at the issues and dilemmas which surround microbiology.

Editor

Contents

1

Origin and Development of Microbiology

The history of microbiology, like all human history, is not a catalog of linear progress, but is more of an interweaving of the careers of bright individuals and their insights. The science of microbiology had its most significant early impact on human health, uncovering the cause of the major killers of the day, and then methods to treat them. As microbiology matured, scientists began to look at what non-pathogenic microbes were doing in the environment and we will look a bit at the history of general microbiology.

From the beginning of microbiology, significant resources have been spent to understand and fight disease-causing microorganisms. You may be surprised to learn that only a small fraction of microbes are involved in disease; many other microbes actually enhance our well-being. The harmless microbes that live in our intestines and on our skin actually help us fight off disease. They actively antagonise other bacteria and take up space, preventing potential pathogens from gaining a foothold on our bodies. Indeed, like all large organisms, humans have entire communities of microorganisms in their digestive systems that contribute to their overall health. The microbial community in humans not only protects us from disease, but also provides needed vitamins, such as B12.

Human health and nutrition also depends on healthy farm animals. Cows, sheep, horses and other ruminant animals utilise their microbial associates to degrade plant material into useful nutrients. Without these bacteria inside ruminants, growth on plant material would be impossible. Commercial crops are also central to human prosperity, and much of agriculture depends upon the activities of microbes. For example, an entire group of plants, the legumes, forms a cooperative relationship with certain bacteria. These bacteria convert nitrogen gas to ammonia for the plant, an important nutrient that is often limiting in the environment.

Microbes also serve as small factories, producing valuable products such as cheese, yogurt, beer, wine, organic acids and many other items. In conclusion, while it is less apparent to us, the positive role of microbes in human health is at least as important as the negative impact of pathogens.

MICROBES AND THE ENVIRONMENT

Microbes are the major actors in the synthesis and degradation of all sorts of important molecules in environments. cyanobacteria and algae in the oceans are responsible for the majority of photosynthesis on Earth. They are the ultimate source of food for most ocean creatures (including whales) and replenish the world's oxygen supply. Cyanobacteria also use carbon dioxide to synthesise all of their biological molecules and thus remove it from the atmosphere. Since carbon dioxide is a major greenhouse gas, its removal by cyanobacteria affects the global carbon dioxide balance and may be an important mitigating factor in global warming.

In all habitats, microorganisms make nutrients available for the future growth of other living things by degrading dead organisms. Microbes are also essential in treating the large volume of sewage and wastewater produced by metropolitan areas, recycling it into clean water that can be safely discharged into the environment.

Less helpfully (from the view of most humans), termites contain microorganisms in their guts that assist in the digestion of wood, allowing the termites to extract nutrients from what would otherwise be indigestible. Understanding of these systems helps us to manage them responsibly and as we learn more we will become ever more effective stewards.

Energy is essential for our industrial society and microbes are important players in its production. A significant portion of natural gas comes from the past action of methanogens (methane-producing bacteria). Numerous bacteria are also capable of rapidly degrading oil in the presence of air and special precautions have to be taken during the drilling, transport and storage of oil to minimise their impact.

In the future, microbes may find utility in the direct production of energy. For example, many landfills and sewage treatment plants capture the methane produced by methanogens to power turbines that produce electricity. Excess grain, crop waste and animal waste can be used as nutrients for microbes that ferment this biomass into methanol or ethanol. These biofuels are presently added to gasoline and thus decreasing pollution. They may one-day power fuel cells in our cars, causing little pollution and having water as their only emission.

The history of microbiology, like all human history, is not a catalog of linear progress, but is more of an interweaving of the careers of bright individuals and their insights. Each new discovery relied on previous ones and in turn spawned further inquiry. A web of interdependent concepts evolved over time through the work of scientists in many related disciplines and nations. Often the research of one individual impacted the efforts of another studying a completely different problem.

The science of microbiology had its most significant early impact on human health, uncovering the cause of the major killers of the day, and then methods to treat them. As microbiology matured, scientists began to look at what non-pathogenic microbes were doing in the environment and we will look a bit at the history of general microbiology.

DISCOVERY OF MICROBES

For years the existence of microorganisms was suspected, but could not be proven, since bacteria were too small to be seen with the naked eye. It took the microscope to expose their tiny world and that instrument has been linked to microbiology ever since. In 1664, Robert Hooke devised a compound microscope and used it to observe fleas, sponges, bird feathers, plants and molds, among other items. His work was published in *Micrographia* and became a popular and widely read book at the time.

Several years later Anton van Leeuwenhoek, a fabric merchant and amateur scientist (or "natural philosopher" as such people called themselves), became very adept at grinding glass lenses to make telescopes and microscopes. While crude by modern standards, his were a technical marvel for the time, able to magnify samples greater than 200-fold. They also produced clearer images than the compound microscopes of the time. By peering through his microscope, Leeuwenhoek observed tiny organisms or "wee animacules" as he called them.

He spent months looking at every kind of sample he could find and eventually submitted his observations in a letter to the Royal Society of London, causing a sensation. Hooke was asked to confirm the findings of Leeuwenhoek and his affirmative assessment garnered them wide acceptance. Surprisingly the work of these two scientists was not followed up for almost 200 years. Human societies had neither the technical prowess nor the inclination to develop the science of microorganisms. It was not until the rise of the industrial revolution that governments and people dedicated the financial and physical resources to understand these small inhabitants of our world.

With the development of better microscopes in the 19th century, scientists returned to an examination of microorganisms. After finishing his education, Ferdinand Julius Cohn was able to convince his father to lay down the large sum necessary to purchase a microscope for him, one better than that available at the University in Breslau, then part of Germany. He used it to carefully examine the world of the microbe and made many observations of eukaryotic microorganisms and bacteria. His landmark papers on the cycling of elements in nature was published in *Ueber Bakterien* in 1872 and a microbial classification scheme including descriptions of *Bacillus* were published in the first volume of a journal he founded, *Beitraege zur Biologie der Planzen*. Cohn's work with microscopes popularised their use in microbiology. This and his other work inspired many other scientists to examine microbes. Cohn's encouragement of Robert Koch, a German physician by training, began the field of medical microbiology.

A major contribution to bacterial techniques was the development of methods using solid medium for the cultivation of bacteria. Robert Koch was convinced that microbes caused some diseases. However, to test this idea, he needed to isolate the causative agent. Almost all samples from diseased animals or any natural surface contained many different microbes and it was impossible to tell which one was the problem. A method was needed to separate these different bacteria. The most common method of isolation was to continually dilute a sample in liquid broth in hopes that at high enough dilution, only one type of microbe would be found. A problem with this method is that only the most populous microbe would be isolated, but that might not be one causing the disease.

There were other technical problems as well with such a liquid-based system, so a solid medium would seem to provide distinct advantages. Koch had tried gelatin for these experiments with unsatisfactory results. Building on the work of Brefeld and Schroeter, Koch used potato slices as a solid medium and observed that a boiled potato left in the open air would develop tiny circular raised spots.

Examination of these spots revealed they were made up of microorganisms and each spot had just one type of microbe in it. He realised that these colonies were pure cultures of bacteria and probably arose from a single species of microbe from the air that landed on the potato. By boiling a potato, slicing it with a hot knife and keeping it in a sterile container with a lid, Koch could keep the potato sterile. But if a sample from a disease animal was smeared across the potato, colonies arose, each being pure isolates from the animal. By then testing these isolates in animals, Koch was able to isolate the cause of anthrax, *Bacillus anthracis*.

Potatoes failed to support the growth of many microorganisms and Koch and his laboratory were constantly frustrated by the lack of a good solid medium. Walter Hesse joined Koch's laboratory to do studies on air quality, showing a remarkable attention to detail and patience in his work. His wife, Angelina Fannie Hesse along with raising their three sons, also would assist her husband with his research in the laboratory. Walter was attempting to do his air quality experiments using medium containing gelatin as the solidifying agent. In the summertime, temperatures would often rise above the melting point of gelatin, then used as the solidifying agent.

In addition, microbes would often grow in the cultures that were capable of degrading gelatin and in both cases this would cause liquefaction of the medium, ruining the experiments. One day while eating lunch, the frustrated scientist asked Lina (as she was called) why her jellies and puddings stayed solid even in the hot summer temperatures. She told him about agar-agar, a heat resistant gelling agent that she had learned about while growing up in New York from a Dutch neighbor who had emigrated from Java.

Development of the new agent by Angelina and Walter led to a resounding success. Few microbes are able to degrade agar and it melts at 100°C yet remains molten at temperatures above 45°C. This allows the mixing of the agar with heat-sensitive nutrients and microbes. After solidification, it will not melt until a temperature of 100 °C is again attained, facilitating the easy cultivation of pathogens. It can also be stored for long periods of time, allowing the cultivation of slow-growing microorganisms. Any type of broth can be mixed with agar, giving great flexibility in the kinds of medium that can be made. Thus, many more types of microbes could be cultivated.

Koch's laboratory also developed methods of pure culture maintenance and aseptic technique. Aseptic technique involves the manipulation of pure cultures in a manner that prevents their contamination by outside microorganisms. Equally important, aseptic technique prevents their spread into the environment. Remember that Koch was studying some of the most devastating microbial pathogens of the period, and their release could potentially cause disease in the scientists working on them. These procedures were also absolutely critical because they allowed careful study of pure microorganisms, making it possible to identify the role of each microbe in a given situation.

Another problem in the cultivation of microbes was solved by Julius Petri while working in Koch's laboratory. Solid medium was poured on glass plates and allowed to spread and harden. Once cooled it allowed a solid surface for streaking. However, creation of these plates required great care since exposure to the air (and the microbes in it) often lead to contamination. In addition, to prevent contamination of plates during incubation, a cumbersome bell jar was used. If one wanted to view samples, the plate had to be removed from the jar, further exposing it to unwanted microbes in the air. In 1887 Petri developed shallow glass dishes, with one having a slightly larger diameter than the other. Medium is poured into the smaller dish and the larger one serves as a cover. This simple device solved all of the above problems and took on the name of its inventor, the petri plate.

These same techniques are essential in studying all microorganisms. Collectively the above techniques have been used to isolate and identify thousands of different microorganisms. As a testament to the significance of their achievement, these techniques are practiced with remarkably little change in every laboratory that works with microorganisms today.

Spontaneous generation was a widely held belief throughout the middle ages and into the latter half of the 19th century. In fact, some people still believe in it today. The idea was attractive because it meshed nicely with the prevailing religious views of how God created the universe. There was a strong bias to legitimise the idea because this vital force was considered a strong proof of God's presence in the world. Many recipes and experiments were offered in proof. To create mice, a recipe called for dirty underwear and wheat grain to be mixed in a bucket and left open outside. In 21 days or less, you would have mice. The real cause may seem obvious from a modern perspective, but to the proponents of this idea, the mice spontaneously arose from the wheat kernels.

Another often-used example was the generation of maggots from meat that was left in the open. The failing here was revealed by Francesco Redi in 1668 with a classic experiment. Redi suspected that flies landing on the meat laid eggs that eventually grew into maggots. Here he used three pieces of meat. One piece of meat was placed under a piece of paper. The flies could not lay eggs onto the meat and no maggots developed. The second piece was left in the open air, resulting in maggots. In the final test, a third piece of meat was overlaid with cheesecloth. The flies were able to lay the eggs into the cheesecloth and when this was removed no maggots developed.

However, if the cheesecloth containing the eggs was placed on a fresh piece of meat, maggots developed, showing it was the eggs that "caused" flies and not spontaneous generation. This helped to end the debate about spontaneous generation for large organisms. However, spontaneous generation was so seductive a concept that even Redi believed it was possible in other circumstances.

DEVELOPMENT OF VACCINATION

Smallpox was a feared disease throughout human history and justifiably so. It was highly contagious and almost everyone eventually became infected. Mortality rates were as high as 25% in adults and closer to 40% in children. Those who did survive often had scarring due to the blister-like pustules that form on the skin, but they obtained life-long immunity to the disease.

As far back at the 11th century in India and China it was realised that liquid from the pustules of a smallpox victim, when scratched on the skin of a healthy patient, would most often cause mild disease. This intentional infection, termed variolation, would also give life-long protection against the virus. Lady Mary Wortley Montgue, wife of ambassador to the Ottoman Empire, introduced variolation to England in 1721 and it became a popular practice throughout Europe. Washington even began variolating the Continental Army in 1776.

Variolation had some deleterious side effects. Serious skin lesions inevitably resulted at the site of inoculation, often accompanied by a generalised rash or even a full case of smallpox. The fatality rate from variolation was 1 to 2 %.

Today we would find this level of fatality to be unacceptable, but at the time this risk still represented a significant advance.

In 1796, Edward Jenner, went in search of a more predictable and safer method of protection against the disease. He noticed that milkmaids rarely contracted smallpox. Further investigation revealed they often contracted cowpox from their charges. Jenner hypothesised that cowpox was related to smallpox and contraction of the former would protect against the latter. In a classic experiment (and one that would land you in jail today), Jenner inoculated a young patient with cowpox and later challenged him with smallpox. The boy did not become ill and Jenner was responsible for the creation of a safer method of protection against smallpox. It is important to stress that the nature of these diseases and their viruses would not be known for over 100 years. Jenner was ahead of his time.

Beginning around 1876 Pasteur's studies on chicken cholera led to the development of vaccines to fight the disease. Cholera was a serious problem since it was able to spread through a barnyard and wipe out a flock in as little as 3 days. It was transmitted by contaminated food or animal excrement. Pasteur identified the cholera bacillus and grew it in pure culture. When injected with it, a chicken invariably died within 48 hours.

Then, as often happens in scientific research, luck intervened. During the heat of the summer, Pasteur returned to Paris and left the cholera cultures used for infection stored on the shelves of his laboratory in Arbois, France. Upon returning, something had happened to the cultures, they no longer caused disease when tested in chickens. With some impatience for the time they were wasting, his group set to work making new cultures of the bacillus and tested these batches on both new birds and also those previously inoculated with the ineffective strain. To their amazement the previously injected birds were unaffected by the fresh bacillus culture, while the new birds all died. Pasteur immediately realised that this was similar to the studies of Jenner.

Pasteur then developed a method for creating cultures that would confer immunity, but not cause disease. Sometimes this involved growing the microbe in medium in the laboratory where they would spontaneously loose their virulence. In other cases it involved multiple passes through a susceptible host. For example, the rabies virus was attenuated by passing it through rabbits.

In honor of Jenner's accomplishments Pasteur coined the term vaccination for the process of immunisation against disease. In this and several of Pasteur's other discoveries, luck played a part, but it was only helpful because he tenaciously pursued "odd" results and had the insight to arrive at important conclusions. In Pasteur's famous words, " In the field of observation, chance favors only the prepared mind."

Pasteur's technique of weakening a strain by a damaging treatment or passing it through a susceptible host was termed attenuation and resulted in the creation of vaccines against anthrax, plague, yellow fever, rabies and many other diseases. Many vaccines have been developed over the years and children today receive a number of shots, greatly decreasing infant mortality.

ANTIMICROBIAL COMPOUNDS

As scientists manipulated these microbes in the lab they found that certain dyes and other compounds were able to inhibit their growth. This inspired Paul Ehrlich to propose that chemicals may exist that will kill the microbe, but not the patient. The hope was that the chemicals will cure the illness of a patient. One of the diseases Ehrlich hoped to cure was syphilis, which had reached epidemic proportions in Europe. Little did he realise it would be a 17-year odyssey before he would develop salvarsan, the first effective chemotherapeutic agent. Salvarsan was the 606th chemical he tried. This arsenic compound effectively kills *Treponema pallidum*, the causative agent of syphilis.

The treatment, however, had many problems, causing long lasting health complications for those individuals who used it. In addition, despite *Treponema* being quite sensitive to salvarsan, the physician had to administer it intravenously for optimum effectiveness. Intravenous injection was a recent development and many doctors were leery of trying the procedure. In London, a young physician by the name of Alexander Fleming, then in the Army Medical Corps, was one of the few that was willing to treat patients. Fleming even got the nickname private 606 from his burgeoning practice. His work validated the effectiveness of salvarsan against syphilis and convinced others to administer the treatment.

Fleming, was a physician by training, but spent most of his time studying bacteria, and his success with salvarsan motivated him to search for other antibacterial agents. His first discovery was lysozyme, an enzyme produced by many organisms, including humans, that lysed some bacteria. This enzyme is not useful as a therapeutic agent because it is difficult to administer as a drug, but Fleming did develop titration methods and assays that would become very useful.

Fleming's arguably most important contribution to science is his discovery of penicillin. In September of 1928, before leaving on a summer holiday, Fleming streaked some plates of *Staphylococcus aureus* and left them to incubate until his return. In an improbable set of circumstances, the beginning of the holiday was cold, allowing some contaminating mold spores (that had blown in from a nearby window) to grow up on some of the plates. The temperature then increased encouraging the growth of the *Staphyloccoccus*.

Many experimenters when confronted with a contaminated plate will look for the trash bin, but Fleming instead spent some time examining it. The fungus had a zone of clearing around it where the *Staphylococcus* colonies would not grow, suggesting the fungus was producing an antibacterial compound that had diffused into the medium. Intrigued, he cultured the fungus, a *Penicillium* mold, and eventually isolated a soluble extract that could kill bacteria and treat localised infection. He called the new compound penicillin after the mold from which it came. Due to the technology then available, however, it was very difficult to prepare a solution that could be used throughout the body without causing problems.

World War II added a greater urgency to the search for compounds that could fight infectious disease. Wounded soldiers, if they survived the initial injury, would often develop life-threatening infections and there were no effective drugs to combat them. In 1939 Howard Florey and Ernst Chain began a systematic study of antimicrobial compounds in hopes of developing treatments for these soldiers and ran across Fleming's report written 9 years earlier. They now were able to purify the compound completely and describe its high potency against microbes.

The availability of penicillin during World War II saved countless lives. The rediscovery of penicillin touched off a search for other microbes producing substances that could kill or inhibit microbes, leading to the discovery of many more antimicrobials. In the ensuing decades the search and discovery of numerous antimicrobial compounds, combined with the development of vaccines, has eliminated many of the deadly diseases that plagued humankind. While we now realise this is a continuing war and not a one-time battle, our understanding of these microbes and the nature of disease will likely keep infectious disease at bay for the foreseeable future.

Understandably, early work in microbiology focused on pathogenic microbes and treatment of the diseases they caused. Martinus Beijerinck and Sergei Winogradsky began the transition from this early work into the present era where we study the molecular biology of a wide variety of microorganisms. Both were interested in microbes present in the soil and water and their work founded the discipline of environmental microbiology. Much of what we understand about bacteria in the environment and their impact can be traced back to the efforts of these two scientists.

Martinus Beijerinck, was originally trained as a botonist and began his work studying the microorganisms that were present in and around plants. He soon began experiments with microbes in the soil. His greatest contribution was the development of enrichment media. Previously, microbes were cultivated on medium consisting of potatoes or extracts of left-over animal renderings. Such media would support the growth of many different bacteria, with chance and population density dictating what became dominant in the culture. In many cases this was exactly what was desired, but in other cases it was of interest to find bacteria capable of performing certain chemical conversions.

Beijerinck discovered that by adding or removing certain compounds from the medium or incubating under different conditions, it was possible to favor the growth of certain microbes and prevent the growth of others. An example is Beijerinck's work with nitrogen-fixing bacteria, which are important in agriculture and the global cycling of nitrogen. They are capable of taking nitrogen gas (N_2) from the air and reducing it to ammonia (NH_3).

This is an important property since most organisms, including agriculturally important plants, can only use reduced nitrogen compounds such as ammonia. Beijerinck wanted to isolate an organism capable of fixing nitrogen in the presence of air, because all previous isolates fixed nitrogen only under anaerobic conditions. By making medium that did not contain a source of fixed nitrogen and then incubating in the presence of air (containing N_2), he demanded that any microbe growing in the medium had to be able to derive its nitrogen by performing aerobic nitrogen fixation. Using this method he succeeded in isolating a new microbe (*Azotobacter chroococum*) with these capabilities. Similar selective culture techniques (as he liked to call them) enabled his laboratory to isolate sulfur-reducing and sulfur-oxidizing bacteria, *Lactobacillus* species, green algae and many other microbes. Selective culture techniques have since been applied to many different groups of microorganisms, allowing them to easily be brought into pure culture.

Sergei Winogradsky, was also interested in soil bacteria, especially those involved in the cycling of nitrogen and sulfur compounds. He was one of the first to isolate microorganisms responsible for the conversion of these elements in the soil, obtaining pure cultures of bacteria capable of the conversion of ammonia to nitrate by microorganisms in the soil. Winogradsky also studied the consumption of hydrogen sulfide gas by sulfur-oxidizing bacteria directly in their natural habitat.

When working with *Beggiatoa* (one of these sulfur-oxidizing bacteria) he discovered it was obtaining its cell carbon from carbon dioxide. He used the term autotrophy to describe this property, a radical idea at the time, and one not readily accepted by other microbiologists. This led Winogradsky to propose the concept of chemolithotrophy; bacteria capable of growth using purely inorganic sources of carbon and energy. He realised that these microbes were essential for the cycling of nutrients in the environment, and the living world was dependent upon their action, long before others accepted the idea.

As a result of the work of Winogradsky and Beijerinck there was great enthusiasm for identifying and classifying the bacteria inhabiting our natural world. For the first part of the 20th century many scientists isolated microbes and made proposals for their organisation into genera and species. In 1909 Sigurd Orla-Jensen suggested a classification scheme based on the functions and abilities of the bacteria, such as growth on certain compounds or the production of specific by-products. The Society of American Bacteriologists, later to become the American Society of Microbiology, applied this technique to prepare a report on the classification of bacteria. This eventually evolved into Bergey's Manual of Determinative Bacteriology in 1923 and subsequent editions have become authoritative reference works.

There was great optimism that as more bacteria were brought into pure culture a clear organisation, based on the evolutionary history of microbes, would emerge from studies of their physiology. However, by the 1940's it became clear that bacteria were unwilling to go along with this idea because classification based on one set of properties was often inconsistent with classification using a different set of properties. Scientists threw up their hands and gave up the idea of finding a system of classification based on growth and morphological characteristics.

Several decades passed during which bacteria became tools for understanding life, but their place in evolution was of little interest. Then in 1961 Brian McCarthy and E. T. Bolton developed methods for the comparison of genetic material between species and Linus Pauling and Emile Zuckerkandl formalised the idea of using the makeup of biological molecules, their sequence, as a way to determine phylogenetic relationships.

Later, Carl Woese, began to use this insight, choosing the sequence of the 16S rRNA from the ribosomes of bacteria as basis for evolutionary comparisons. This analysis was a watershed event for the evolutionary classification of bacteria and in 1977 Woese used this technique to assert that known bacteria should be divided into two separate domains that we now call Bacteria and Archaea, radically changing our view of the microbial world.

Lynn Margulis proposed the remarkable notion that some bacteria fused with archaea to create eukaryotes. Endosymbiosis, has been strongly supported by a range of subsequent biochemical data. In subsequent years, determination of 16S rRNA sequences, and then complete genomes, of microorganisms has enabled the natural classification of microorganisms.

GENOME DEVELOPMENT

Genomic sequences of organisms, including humans and many important pathogens, are generating new insights into us, as well as the diseases caused by certain microbes. This large amount of data has spawned the fields of genomics and bioinformatics.

One final thread we will trace is the rise of molecular microbiology and the technique of genomics. This pathway begins in 1885 with the seemingly unimportant isolation of a common intestinal microorganism by Theodor Escherich. The microbe is eventually renamed *Escherichia coli* in his honor. It could not have been understood at the time, but this bacterium would play a central role in our understanding of life. *E. coli* achieved this position because it has several properties that made it invaluable as an experimental system:

E. coli is easy to take care of. It can grow on a wide variety of media, at temperatures ranging from 20-45 °C, in the presence or absence of air. *E. coli* grows quickly. Under normal laboratory conditions it will go through a complete life cycle, doubling its population about every 30 minutes.

With this rapid growth rate it is possible to generate large populations in just a few hours. Most experiments can be set up one day and the results observed the next.

It can survive storage for long periods of time without losing viability using relatively simple preservation techniques. It was available. There is an element of arbitrariness that this microbe came to serve such a vital role in biology. Many other bacteria probably could have filled this position, but *E. coli* was the one chosen.

Several key discoveries laid the foundation for future developments in molecular biology. The first of these was the discovery of autonomously replicated pieces of DNA separate from the bacterial chromosome. plasmids, as Joshua Lederberg called them, were capable of carrying genes that could be regulated and moved from one microbe to another independently of the bacterial chromosome. The ability of some plasmids to move between different cells meant that different bacteria could rapidly inherit genes carried on these plasmids. It was found that one class of plasmids encoded resistance genes that would enable the bacteria to grow in the presence of an antibiotic. It later became clear it was possible to select for the presence of these plasmids by demanding growth of the bacterial strain on medium containing the antibiotic to which the plasmid encoded resistance.

A second discovery that greatly impacted modern genetic engineering was that bacteria have their own parasites. As mentioned earlier, there are viruses that infect and kill bacteria, just as viruses infect animals and plants. The analysis of bacterial viruses was a major avenue to understanding the bacteria themselves.

For example, investigation of the ability of some viruses to grow well on certain strains of bacteria and not others lead to the discovery of restriction-modification systems in bacteria. These systems allow bacteria to recognise and destroy foreign DNA and consist of two parts. A modification enzyme that labels DNA as belonging to the bacterium and a restriction enzyme that recognises a certain 4-8 base pair sequence in the DNA and cuts it in two if it has not been modified. The landmark paper of Hamilton Smith and Kent Wilcox showed that restriction enzymes were capable of cleaving double-stranded DNA into discrete pieces. The authors correctly hypothesised that the enzyme was recognising a sequence of DNA and cutting it.

Further studies on restriction enzymes led to the birth of genetic engineering. Joan Mertz and Ronald Davis made restriction enzymes into tools for the manipulation of DNA when they showed that many of the breaks these enzymes made in the DNA produced single-stranded complementary ends. Stanley Cohen, Annie Chang, Robert Helling and Herbert Boyer then demonstrated that any DNA can be broken into fragments with restriction enzymes and by mixing that with a plasmid digested in a similar manner, it was possible to create recombinant molecules. Moving the plasmid into a microbe and growing it on selective medium could produce any desired amount of these recombinant molecules.

Another discovery was "brewing" in the western United States. In the mid 1960s, Thomas Brock was on a trip to Yellowstone and became intrigued by the mats of green, brown and pink material that were found in many of the hot springs. Brock was sure that these were living communities of microorganisms, yet the springs were at temperatures near boiling. Subsequent research proved his hypothesis correct and lead to the isolation of a microbe, *Thermus aquaticus*, capable of growth at 85°C.

Organisms capable of growing at this high a temperature were unheard of at the time and the discovery spawned two major developments. First, work with *T. aquaticus* and other unusual microbes by Woese and many others lead to the discovery of the Archaea. Second, the enzyme responsible for replicating DNA in *T. aquaticus* was found to be extremely heat stable, surviving temperatures of 95°C and copying DNA at 72°C.

The use of *T. aquaticus* polymerase sped the adoption of polymerase chain reaction (always referred to as PCR). This technique was developed in 1985 by Kary Mullis, and became incredibly useful for amplifying even small quantities of DNA. Originally the technique required fresh addition of polymerase enzyme after each step, which could become prohibitively expensive. The use of *T. aquaticus* polymerase made the procedure much more economical. PCR has become a pivotal technique in many areas of science, from detecting bacteria in the environment to the analysis of evidence at crime scenes.

In 1977 Walter Gilbert and Fred Sanger independently developed methods for determining the exact sequence of bases in DNA. These techniques became immediately useful in determining the sequence of numerous important genes being investigated in countless laboratories.

It became efficient enough that in 1985 Robert Sinsheimer convened a meeting of a number of biologists active in genetics and gene mapping. Out of this meeting came a proposal to sequence the entire three billion base pairs of the human genome. This was equivalent at the time of proposing a mission to the moon, technically possible, but with unpredictable value. The initial cost of the project was estimated to be $10 a base pair or a staggering 30 billion dollars. There was resistance from a significant portion of the scientific community, fearing that the project would siphon funds away from other worthy scientific projects. However, the clarity of the goal ignited the imagination of Congress and the public, creating unstoppable momentum for the project. After about a decade of effort, a preliminary draft of the entire genome was released in the spring of 2001, earlier than projected and dramatically under budget. The project itself drove the development of sequencing technology, allowing the determination of millions of base pairs of sequence per day at a cost of less than 3 cents per base at large sequencing facilities.

Along with the human genome, many other model organisms have been or are about to be sequenced. These include the genomes of many pathogenic bacteria, the mouse, the fruit fly and the nematode *Caenorhabditis elegans*, the last being an experimental model for development in eukaryotes.

TODAY'S MICROBIOLOGY

Looking back over the emergence of molecular genetics during the past 50 years, from the deciphering of the genetic code to the advent of recomb-inant DNA, the recognition of the cell cycle and the latest developments in intra- and intercellular signalling, it's clear that bacteria and viruses (especially phages) have been major players at every stage. Genetically modified interferons, insulin and other drugs, which are now saving countless lives and ameliorating many more, comprise just one practical outcome of the knowledge gained from this great collaboration between microbes and humans. Bacteria are also moving rapidly into clinical practice in their own right .

One that has entered clinical trials recently is Streptococcus lactis, genetically modified not to make enamel-eroding lactic acid, which is introduced into the mouth to replace those strains that do rot teeth. Others include a Lactobacillus lactis recombinant that secretes a therapeutic protein to help Crohn's disease patients; and a Lactobacillus crispatus strain to combat recurrent urinary tract infection. Meanwhile Clostridium novyi, an anaerobe that can infect hypoxic regions within tumours, is showing promise as a means of delivering chemotherapeutic agents.

Reviewing the scientific insights now being provided by micro-organisms in abundance, one is impressed above all by their diversity. A fast-evolving feline immunodeficiency virus, for example, has revealed details of the population structure and demographic history of its natural wildlife host (the cougar) that were impossible to obtain by any other means. Studies on the deep-sea-living Photobacterium profundum have yielded a first glimpse into the molecular basis for life in the largest portion of the biosphere, and single-celled algae have shown how abnormalities in human cilia can explain the basis not only of Bardet-Biedl syndrome but also of many other conditions too.

Whatever the truth behind contemporary cogitations regarding global warming, it's clear that microbial life will be here long after we have departed, just as it flourished on Earth aeons ahead of our arrival. Those verities alone tell us much about the resilience and diversity of micro-organisms.

REFERENCES

Dreyfus, D. H., "Evidence suggesting an evolutionary relationship between transposable elements and immune system recombination sequences", *Molecular Immunology,* 1992.

Farries, T. C., Steuer, K. L. and Atkinson, J. P., "Evolutionary implications of a new bypass activation pathway of the complement system", *Immunology Today,* 1990.

Hansen, J. D. and McBlane, J. F., "Recombination-Activating Genes, Transposition, and the Lymphoid-Specific Combinatorial Immune System: A Common Evolutionary Connection", *Origin and Evolution of the Vertebrate Immune System,* Edited by L. Du Pasquier and G. W. Litman. Berlin, Springer, 2000.

Marchalonis, J. J., Jensen, I. and Schluter, S. F., "Structural, antigenic and evolutionary analyses of immunoglobulins and T cell receptors." *Journal of Molecular Recognition,* 2002

2

Microbial Diversity

The microbial world is the largest unex-plored reservoir of biodiversity on theEarth. It is an important frontier in biologyunder intensive investigations. Interest inthe exploration of microbial diversity has been spurred by the fact that microbes areessential for life since they perform numer-ous functions essential for the biospherethat include nutrient cycling and environ-mental detoxification. The vast array ofmicrobial activities and their importanceto the biosphere and to human economies provide strong rationale for understandingtheir diversity, conservation and exploitationfor society.

MICROORGANISMS

The existence of microorganisms was hypothesised during the late Middle Ages but they were not observed or proven until the invention of the microscope in the 17th century. In The Canon of Medicine, Abu Ali ibn Sina stated that bodily secretion is contaminated by foul foreign earthly bodies before being infected, but he did not view them as primary causes of disease. When the Black Death bubonic plague reached al-Andalus in the 14th century, Ibn Khatima and Ibn al-Khatib hypothesised that infectious diseases are caused by microorganisms which enter the human body. Such ideas became more popular in Europe during the

Renaissance, particularly through the writing of the Italian monk Girolamo Fracastoro.

Prior to Anton van Leeuwenhoek's discovery of microorganisms in 1675, it had been a mystery as to why grapes could be turned into wine, milk into cheese, or why food would spoil. Leeuwenhoek did not make the connection between these processes and microorganisms, but using the microscope, he did establish that there were forms of life that were not visible to the naked eye. Leeuwenhoek's discovery, along with subsequent observations by Lazzaro Spallanzani and Louis Pasteur, ended the long-held belief that life spontaneously appeared from non-living substances during the process of spoilage.

Lazzarro Spallanzani found that microorganisms could only settle in a broth if the broth was exposed to the air. He also found that boiling the broth would sterilise it and kill the microorganisms. Louis Pasteur expanded upon Spallanzani's findings by exposing boiled broths to the air, in vessels that contained a filter to prevent all particles from passing through to the growth medium, and also in vessels with no filter at all, with air being admitted via a curved tube that would not allow dust particles to come in contact with the broth. By boiling the broth beforehand, Pasteur ensured that no microorganisms survived within the broths at the beginning of his experiment. Nothing grew in the broths in the course of Pasteur's experiment. This meant that the living organisms that grew in such broths came from outside, as spores on dust, rather than spontaneously generated within the broth. Thus, Pasteur dealt the death blow to the theory of spontaneous generation and supported germ theory.

In 1876, Robert Koch established that microbes can cause disease. He did this by finding that the blood of cattle who were infected with anthrax always had large numbers of Bacillus anthracis. Koch also found that he could transmit anthrax from one animal to another by taking a small sample of blood from the infected animal and injecting it into a healthy one, causing the healthy animal to become sick. He also found that he could grow the bacteria in a nutrient broth, inject it into a healthy animal, and cause illness. Based upon these experiments, he devised criteria for establishing a causal link between a microbe and a disease in what are now known as Koch's postulates. Though these postulates cannot be applied in all cases, they do retain historical importance in the development of scientific thought and can still be used today.

MICROORGANISMS CLASSIFICATION

Microorganisms can be found almost anywhere in the taxonomic organisation of life on the planet. Bacteria and archaea are almost always microscopic, while a number of eukaryotes are also microscopic, including most protists and a number of fungi. Viruses are generally regarded as not living and therefore are not microbes, although the field of microbiology also encompasses the study of viruses.

Prokaryotes

Prokaryotes are organisms that lack a cell nucleus and the other organelles found in eukaryotes. Prokaryotes are almost always unicellular, although some such as myxobacteria can aggregate into complex structures as part of their life cycle. These organisms are divided into two groups, the archaea and the bacteria.

BACTERIA

Bacteria are unicellular microorganisms. Typically a few micrometres in length, individual bacteria have a wide-range of shapes, ranging from spheres to rods to spirals. Bacteria are ubiquitous in every habitat on Earth, growing in soil, acidic hot springs, radioactive waste, seawater, and deep in the Earth's crust. There are typically 40 million bacterial cells in a gram of soil and a million bacterial cells in a millilitre of fresh water; in all, there are approximately five nonillion (5×1030) bacteria in the world.

Bacteria are vital in recycling nutrients, and many important steps in nutrient cycles depend on bacteria, such as the fixation of nitrogen from the atmosphere. However, most of these bacteria have not been characterised, and only about half of the phyla of bacteria have species that can be cultured in the laboratory. The study of bacteria is known as bacteriology, a branch of microbiology.

There are approximately 10 times as many bacterial cells as human cells in the human body, with large numbers of bacteria on the skin and in the digestive tract. Although the vast majority of these bacteria are rendered harmless or beneficial by the protective effects of the immune system, a few pathogenic bacteria cause infectious diseases, including cholera, syphilis, anthrax, leprosy and bubonic plague.

The most common fatal bacterial diseases are respiratory infections, with tuberculosis alone killing about 2 million people a year, mostly in sub-Saharan Africa. In developed countries, antibiotics are used to treat bacterial infections and in various agricultural processes, so antibiotic resistance is becoming common. In industry, bacteria are important in processes such as wastewater treatment, the production of cheese and yoghurt, and the manufacture of antibiotics and other chemicals.

Bacteria are prokaryotes. Unlike cells of animals and other eukaryotes, bacterial cells do not contain a nucleus and rarely harbour membrane-bound organelles. Although the term bacteria traditionally included all prokaryotes, the scientific classification changed after the discovery in the 1990s that prokaryotic life consists of two very different groups of organisms that evolved independently from an ancient common ancestor. These evolutionary domains are called Bacteria and Archaea.

Bacteriology

The existence of microorganisms was hypothesised during the late Middle Ages. In The Canon of Medicine (1020), Abu Ali ibn Sina (Avicenna) stated that bodily secretions are contaminated by "foul foreign earthly bodies" before a person becomes infected, but he did not view these bodies as primary causes of disease. When the Black Death bubonic plague reached al-Andalus in the 14th century, Ibn Khatima and Ibn al-Khatib wrote of infectious diseases being caused contagious entities that enter the human body. These ideas about the contagious nature of some diseases became more popular in Europe during the Renaissance, particularly through the writing of Girolamo Fracastoro.

Bacteria were first observed by Anton van Leeuwenhoek in 1676, using a single-lens microscope of his own design. He called them "animalcules" and published his observations in a series of letters to the Royal Society. Louis Pasteur demonstrated in 1859 that the fermentation process is caused by the growth of microorganisms, and that this growth is not due to spontaneous generation. Along with his contemporary, Robert Koch, Pasteur was an early advocate of the germ theory of disease.

Robert Koch was a pioneer in medical microbiology and worked on cholera, anthrax and tuberculosis. In his research into tuberculosis, Koch finally proved the germ theory, for which he was awarded a Nobel Prize in 1905. In Koch's postulates, he set out criteria to test if an organism is the cause of a disease; these postulates are still used today.

Though it was known in the nineteenth century that bacteria are the cause of many diseases, no effective antibacterial treatments were available. In 1910, Paul Ehrlich developed the first antibiotic, by changing dyes that selectively stained Treponema pallidum-the spirochete that causes syphilis-into compounds that selectively killed the pathogen. Ehrlich had been awarded a 1908 Nobel Prize for his work on immunology, and pioneered the use of stains to detect and identify bacteria, with his work being the basis of the Gram stain and the Ziehl-Neelsen stain.

A major step forward in the study of bacteria was the recognition in 1977 by Carl Woese that archaea have a separate line of evolutionary descent from bacteria. This new phylogenetic taxonomy was based on the sequencing of 16S ribosomal RNA, and divided prokaryotes into two evolutionary domains as part of the three-domain system.

Origin and Early Evolution

The ancestors of modern bacteria were single-celled microorganisms that were the first forms of life to develop on earth, about 4 billion years ago. For about 3 billion years, all organisms were microscopic, and bacteria and archaea were the dominant forms of life.

Although bacterial fossils exist, such as stromatolites, their lack of distinctive morphology prevents them from being used to examine the past history of bacterial evolution, or to date the time of origin of a particular bacterial species. However, gene sequences can be used to reconstruct the bacterial phylogeny, and these studies indicate that bacteria diverged first from the archaeal/ eukaryotic lineage. The most recent common ancestor of bacteria and archaea was probably a hyperthermophile that lived about 2.5 billion-3.2 billion years ago.

Bacteria were also involved in the second great evolutionary divergence, that of the archaea and eukaryotes. Here, eukaryotes resulted from ancient bacteria entering into endosymbiotic associations with the ancestors of eukaryotic cells, which were themselves possibly related to the Archaea. This involved the engulfment by proto-eukaryotic cells of alpha-proteobacterial symbionts to form either mitochondria or hydrogenosomes, which are still being found in all known Eukarya.

Later on, an independent second engulfment by some mitochondria-containing eukaryotes of cyanobacterial-like organisms led to the formation of chloroplasts in algae and plants. There are even some algal groups known that clearly originated from subsequent events of endosymbiosis by heterotrophic eukaryotic hosts engulfing a eukaryotic algae that developed into "second-generation" plastids.

Bacteria display a wide diversity of shapes and sizes, called morphologies. Bacterial cells are about 10 times smaller than eukaryotic cells and are typically 0.5-5.0 micrometres in length. However, a few species-for example Thiomargarita namibiensis and Epulopiscium fishelsoni-are up to half a millimetre long and are visible to the unaided eye. Among the smallest bacteria are members of the genus Mycoplasma, which measure only 0.3 micrometres, as small as the largest viruses.

Most bacterial species are either spherical, called cocci or rod-shaped, called bacilli. Some rod-shaped bacteria, called vibrio, are slightly curved or comma-shaped; others, can be spiral-shaped, called spirilla, or tightly coiled, called spirochetes. A small number of species even have tetrahedral or cuboidal shapes.

This wide variety of shapes is determined by the bacterial cell wall and cytoskeleton, and is important because it can influence the ability of bacteria to acquire nutrients, attach to surfaces, swim through liquids and escape predators.

Many bacterial species exist simply as single cells, others associate in characteristic patterns: Neisseria form diploids (pairs), Streptococcus form chains, and Staphylococcus group together in ·"bunch of grapes" clusters. Bacteria can also be elongated to form filaments, for example the Actinobacteria. Filamentous bacteria are often surrounded by a sheath that contains many individual cells; certain types, such as species of the genus Nocardia, even form complex, branched filaments, similar in appearance to fungal mycelia.

Bacteria often attach to surfaces and form dense aggregations called biofilms or microbial mats. These films can range from a few micrometers in thickness to up to half a meter in depth, and may contain multiple species of bacteria, protists and archaea. Bacteria living in biofilms display a complex arrangement of cells and extracellular components, forming secondary structures such as microcolonies, through which there are networks of channels to enable better diffusion of nutrients. In natural environments, such as soil or the surfaces of plants, the majority of bacteria are bound to surfaces in biofilms. Biofilms are also important for chronic bacterial infections and infections of implanted medical devices, as bacteria protected within these structures are much harder to kill than individual bacteria.

Even more complex morphological changes are sometimes possible. For example, when starved of amino acids, Myxobacteria detect surrounding cells in a process known as quorum sensing, migrate towards each other, and aggregate to form fruiting bodies up to 500 micrometres long and containing approximately 100,000 bacterial cells. In these fruiting bodies, the bacteria perform separate tasks; this type of cooperation is a simple type of multicellular organisation. For example, about one in 10 cells migrate to the top of these fruiting bodies and differentiate into a specialised dormant state called myxospores, which are more resistant to desiccation and other adverse environmental conditions than are ordinary cells.

Cellular Structures of Bacterial

Intracellular structures

The bacterial cell is surrounded by a lipid membrane, or cell membrane, which encompasses the contents of the cell and acts as a barrier to hold nutrients, proteins and other essential components of the cytoplasm within the cell. As they are prokaryotes, bacteria do not have membrane-bound organelles in their cytoplasm and thus contain few intracellular structures. They consequently lack a nucleus, mitochondria, chloroplasts and the other organelles present in eukaryotic cells, such as the Golgi apparatus and endoplasmic reticulum.

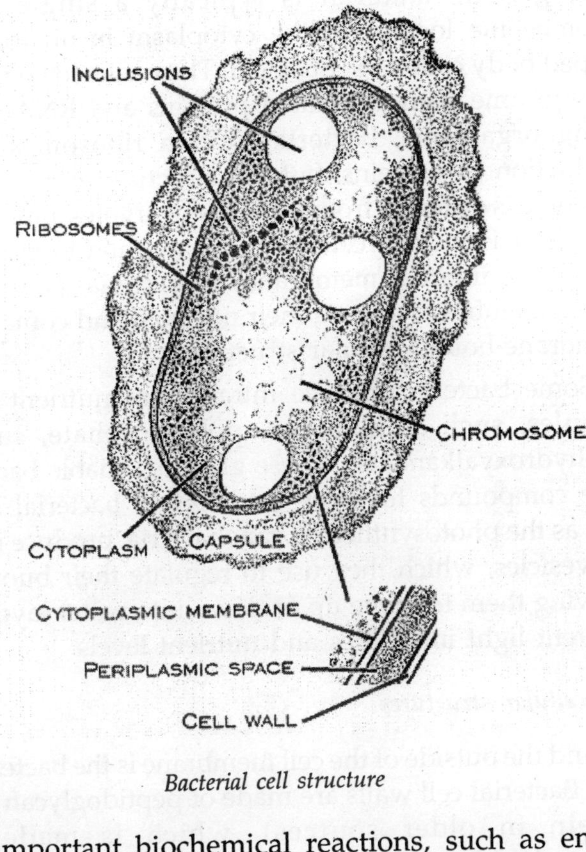

Bacterial cell structure

Many important biochemical reactions, such as energy generation, occur due to concentration gradients across membranes, creating a potential difference analogous to a battery. The absence of internal membranes in bacteria means these reactions, such as electron transport, occur across the cell membrane, between the cytoplasm and the periplasmic space. Additionally, while some transporter proteins consume chemical energy, others harness concentration gradients to import nutrients across the cell membrane or to expel undesired molecules from the cytoplasm.

Bacteria do not have a membrane-bound nucleus, and their genetic material is typically a single circular chromosome located in the cytoplasm in an irregularly shaped body called the nucleoid. The nucleoid contains the chromosome with associated proteins and RNA. Like all living organisms, bacteria contain ribosomes for the production of proteins, but the structure of the bacterial ribosome is different from those of eukaryotes and Archaea. The order Planctomycetes are an exception to the general absence of internal membranes in bacteria, because they have a membrane around their nucleoid and contain other membrane-bound cellular structures.

Some bacteria produce intracellular nutrient storage granules, such as glycogen, polyphosphate, sulfur or polyhydroxyalkanoates. These granules enable bacteria to store compounds for later use. Certain bacterial species, such as the photosynthetic Cyanobacteria, produce internal gas vesicles, which they use to regulate their buoyancy - allowing them to move up or down into water layers with different light intensities and nutrient levels.

Extracellular structures

Around the outside of the cell membrane is the bacterial cell wall. Bacterial cell walls are made of peptidoglycan (called murein in older sources), which is made from polysaccharide chains cross-linked by unusual peptides containing D-amino acids. Bacterial cell walls are different from the cell walls of plants and fungi, which are made of cellulose and chitin, respectively. The cell wall of bacteria is also distinct from that of Archaea, which do not contain peptidoglycan. The cell wall is essential to the survival of many bacteria, and the antibiotic penicillin is able to kill bacteria by inhibiting a step in the synthesis of peptidoglycan.

There are broadly speaking two different types of cell wall in bacteria, called Gram-positive and Gram-negative. The names originate from the reaction of cells to the Gram stain, a test long-employed for the classification of bacterial species.

Gram-positive bacteria possess a thick cell wall containing many layers of peptidoglycan and teichoic acids. In contrast, Gram-negative bacteria have a relatively thin cell wall consisting of a few layers of peptidoglycan surrounded by a second lipid membrane containing lipopolysaccharides and lipoproteins. Most bacteria have the Gram-negative cell wall, and only the Firmicutes and Actinobacteria (previously known as the low G+C and high G+C Gram-positive bacteria, respectively) have the alternative Gram-positive arrangement. These differences in structure can produce differences in antibiotic susceptibility; for instance, vancomycin can kill only Gram-positive bacteria and is ineffective against Gram-negative pathogens, such as Haemophilus influenzae or Pseudomonas aeruginosa.

In many bacteria an S-layer of rigidly arrayed protein molecules covers the outside of the cell. This layer provides chemical and physical protection for the cell surface and can act as a macromolecular diffusion barrier. S-layers have diverse but mostly poorly understood functions, but are known to act as virulence factors in Campylobacter and contain surface enzymes in Bacillus stearothermophilus.

Fimbriae are fine filaments of protein, just 2-10 nanometres in diameter and up to several micrometers in length. They are distributed over the surface of the cell, and resemble fine hairs when seen under the electron microscope. Fimbriae are believed to be involved in attachment to solid surfaces or to other cells and are essential for the virulence of some bacterial pathogens. Pili (sing. pilus) are cellular appendages, slightly larger than fimbriae, that can transfer genetic material between bacterial cells in a process called conjugation.

Capsules or slime layers are produced by many bacteria to surround their cells, and vary in structural complexity: ranging from a disorganised slime layer of extra-cellular polymer, to a highly structured capsule or glycocalyx. These structures can protect cells from engulfment by eukaryotic cells, such as macrophages. They can also act as antigens and be involved in cell recognition, as well as aiding attachment to surfaces and the formation of biofilms.

The assembly of these extracellular structures is dependent on bacterial secretion systems. These transfer proteins from the cytoplasm into the periplasm or into the environment around the cell. Many types of secretion systems are known and these structures are often essential for the virulence of pathogens, so are intensively studied.

Endospores

Certain genera of Gram-positive bacteria, such as Bacillus, Clostridium, Sporohalobacter, Anaerobacter and Heliobacterium, can form highly resistant, dormant structures called endospores.

In almost all cases, one endospore is formed and this is not a reproductive process, although Anaerobacter can make up to seven endospores in a single cell. Endospores have a central core of cytoplasm containing DNA and ribosomes surrounded by a cortex layer and protected by an impermeable and rigid coat.

Endospores show no detectable metabolism and can survive extreme physical and chemical stresses, such as high levels of UV light, gamma radiation, detergents, disinfectants, heat, pressure and desiccation. In this dormant state, these organisms may remain viable for millions of years, and endospores even allow bacteria to survive exposure to the vacuum and radiation in space. Endospore-forming bacteria can also cause disease: for example, anthrax can be contracted by the inhalation of Bacillus anthracis endospores, and contamination of deep puncture wounds with Clostridium tetani endospores causes tetanus.

Microbial Metabolism

In contrast to higher organisms, bacteria exhibit an extremely wide variety of metabolic types. The distribution of metabolic traits within a group of bacteria has traditionally been used to define their taxonomy, but these traits often do not correspond with modern genetic classifications. Bacterial metabolism is classified on the basis of three major criteria: the kind of energy used for growth, the source of carbon, and the electron donors used for growth. An additional criterion of respiratory microorganisms are the electron acceptors used for aerobic or anaerobic respiration.

Carbon metabolism in bacteria is either heterotrophic, where organic carbon compounds are used as carbon sources, or autotrophic, meaning that cellular carbon is obtained by fixing carbon dioxide. Typical autotrophic bacteria are phototrophic cyanobacteria, green sulfur-bacteria and some purple bacteria, but also many chemolithotrophic species, e. g. nitrifying or sulfur-oxidising bacteria. Energy metabolism of bacteria is either based on phototrophy, the use of light through photosynthesis, or on chemotrophy, the use of chemical substances for energy, which are mostly oxidised at the expense of oxygen or alternative electron acceptors.

Finally, bacteria are further divided into lithotrophs that use inorganic electron donors and organotrophs that use organic compounds as electron donors. Chemotrophic organisms use the respective electron donors for energy conservation (by aerobic/anaerobic respiration or fermentation) and biosynthetic reactions, whereas phototrophic organisms use them only for biosynthetic purposes. Respiratory organisms use chemical compounds as a source of energy by taking electrons from the reduced substrate and transferring them to a terminal electron acceptor in a redox reaction.

This reaction releases energy that can be used to synthesise ATP and drive metabolism. In aerobic organisms, oxygen is used as the electron acceptor. In anaerobic organisms other inorganic compounds, such as nitrate, sulfate or carbon dioxide are used as electron acceptors. This leads to the ecologically important processes of denitrification, sulfate reduction and acetogenesis, respectively.

Another way of life of chemotrophs in the absence of possible electron acceptors is fermentation, where the electrons taken from the reduced substrates are transferred to oxidised intermediates to generate reduced fermentation products. Fermentation is possible, because the energy content of the substrates is higher than that of the products, which allows the organisms to synthesise ATP and drive their metabolism.

These processes are also important in biological responses to pollution; for example, sulfate-reducing bacteria are largely responsible for the production of the highly toxic forms of mercury (methyl- and dimethylmercury) in the environment. Non-respiratory anaerobes use fermentation to generate energy and reducing power, secreting metabolic by-products (such as ethanol in brewing) as waste. Facultative anaerobes can switch between fermentation and different terminal electron acceptors depending on the environmental conditions in which they find themselves.

Lithotrophic bacteria can use inorganic compounds as a source of energy. Common inorganic electron donors are hydrogen, carbon monoxide, ammonia (leading to nitrification), ferrous iron and other reduced metal ions, and several reduced sulfur compounds. Unusually, the gas methane can be used by methanotrophic bacteria as both a source of electrons and a substrate for carbon anabolism. In both aerobic phototrophy and chemolithotrophy, oxygen is used as a terminal electron acceptor, while under anaerobic conditions inorganic compounds are used instead. Most lithotrophic organisms are autotrophic, whereas organotrophic organisms are heterotrophic.

Bacterial Growth and Reproduction

Unlike multicellular organisms, increases in the size of bacteria (cell growth) and their reproduction by cell division are tightly linked in unicellular organisms. Bacteria grow to a fixed size and then reproduce through binary fission, a form of asexual reproduction. Under optimal conditions, bacteria can grow and divide extremely rapidly, and bacterial populations can double as quickly as every 9.8 minutes. In cell division, two identical clone daughter cells are produced.

Some bacteria, while still reproducing asexually, form more complex reproductive structures that facilitate the dispersal of the newly formed daughter cells. Examples include fruiting body formation by Myxobacteria and arial hyphae formation by Streptomyces, or budding. Budding involves a cell forming a protrusion that breaks away and produces a daughter cell.

In the laboratory, bacteria are usually grown using solid or liquid media. Solid growth media such as agar plates are used to isolate pure cultures of a bacterial strain. However, liquid growth media are used when measurement of growth or large volumes of cells are required. Growth in stirred liquid media occurs as an even cell suspension, making the cultures easy to divide and transfer, although isolating single bacteria from liquid media is difficult. The use of selective media can help identify specific organisms.

Most laboratory techniques for growing bacteria use high levels of nutrients to produce large amounts of cells cheaply and quickly. However, in natural environments nutrients are limited, meaning that bacteria cannot continue to reproduce indefinitely. This nutrient limitation has led the evolution of different growth strategies.

Some organisms can grow extremely rapidly when nutrients become available, such as the formation of algal (and cyanobacterial) blooms that often occur in lakes during the summer. Other organisms have adaptations to harsh environments, such as the production of multiple antibiotics by Streptomyces that inhibit the growth of competing microorganisms. In nature, many organisms live in communities which may allow for increased supply of nutrients and protection from environmental stresses. These relationships can be essential for growth of a particular organism or group of organisms.

Bacterial growth follows three phases. When a population of bacteria first enter a high-nutrient environment that allows growth, the cells need to adapt to their new environment. The first phase of growth is the lag phase, a period of slow growth when the cells are adapting to fast growth. The lag phase has high biosynthesis rates, as enzymes and nutrient transporters are produced. The second phase of growth is the logarithmic phase (log phase), also known as the exponential phase.

The log phase is marked by rapid exponential growth. The rate at which cells grow during this phase is known as the growth rate (k), and the time it takes the cells to double is known as the generation time (g). During log phase, nutrients are metabolised at maximum speed until one of the nutrients is depleted and starts limiting growth.

The final phase of growth is the stationary phase and is caused by depleted nutrients. The cells reduce their metabolic activity and consume nonessential cellular proteins. The stationary phase is a transition from rapid growth to a stress response state and there is increased expression of genes involved in DNA repair, antioxidant metabolism and nutrient transport.

Genetic Changes

Most bacteria have a single circular chromosome that can range in size from only 160,000 base pairs in the endosymbiotic bacteria Candidatus Carsonella ruddii, to 12,200,000 base pairs in the soil-dwelling bacteria Sorangium cellulosum. Spirochaetes of the genus Borrelia are a notable exception to this arrangement, with bacteria such as Borrelia burgdorferi, the cause of Lyme disease, containing a single linear chromosome. Bacteria may also contain plasmids, which are small extra-chromosomal DNAs that may contain genes for antibiotic resistance or virulence factors.

Another type of bacterial DNA are integrated viruses (bacteriophages). Many types of bacteriophage exist, some simply infect and lyse their host bacteria, while others insert into the bacterial chromosome. A bacteriophage can contain genes that contribute to its host's phenotype: for example, in the evolution of Escherichia coli O157:H7 and Clostridium botulinum, the toxin genes in an integrated phage converted a harmless ancestral bacteria into a lethal pathogen.

Bacteria, as asexual organisms, inherit identical copies of their parent's genes. However, all bacteria can evolve by selection on changes to their genetic material DNA caused by genetic recombination or mutations. Mutations come from errors made during the replication of DNA or from exposure to mutagens. Mutation rates vary widely among different species of bacteria and even among different clones of a single species of bacteria.

Genetic changes in bacterial genomes come from either random mutation during replication or "stress-directed mutation", where genes involved in a particular growth-limiting process have an increased mutation rate. Some bacteria also transfer genetic material between cells. This can occur in three main ways. Firstly, bacteria can take up exogenous DNA from their environment, in a process called transformation. Genes can also be transferred by the process of transduction, when the integration of a bacteriophage introduces foreign DNA into the chromosome.

The third method of gene transfer is bacterial conjugation, where DNA is transferred through direct cell contact. This gene acquisition from other bacteria or the environment is called horizontal gene transfer and may be common under natural conditions. Gene transfer is particularly important in antibiotic resistance as it allows the rapid transfer of resistance genes between different pathogens.

Movement

Motile bacteria can move using flagella, bacterial gliding, twitching motility or changes of buoyancy. In twitching motility, bacterial use their type IV pili as a grappling hook, repeatedly extending it, anchoring it and then retracting it with remarkable force (>80 pN). Bacterial species differ in the number and arrangement of flagella on their surface; some have a single flagellum (monotrichous), a flagellum at each end (amphitrichous), clusters of flagella at the poles of the cell (lophotrichous), while others have flagella distributed over the entire surface of the cell (peritrichous).

The bacterial flagella is the best-understood motility structure in any organism and is made of about 20 proteins, with approximately another 30 proteins required for its regulation and assembly.

The flagellum is a rotating structure driven by a motor at the base that uses the electrochemical gradient across the membrane for power. This motor drives the motion of the filament, which acts as a propeller. Many bacteria have two distinct modes of movement: forward movement (swimming) and tumbling. The tumbling allows them to reorient and makes their movement a three-dimensional random walk. The flagella of a unique group of bacteria, the spirochaetes, are found between two membranes in the periplasmic space. They have a distinctive helical body that twists about as it moves.

Motile bacteria are attracted or repelled by certain stimuli in behaviours called taxes: these include chemotaxis, phototaxis and magnetotaxis. In one peculiar group, the myxobacteria, individual bacteria move together to form waves of cells that then differentiate to form fruiting bodies containing spores. The myxobacteria move only when on solid surfaces, unlike E. coli which is motile in liquid or solid media.

Several Listeria and Shigella species move inside host cells by usurping the cytoskeleton, which is normally used to move organelles inside the cell. By promoting actin polymerisation at one pole of their cells, they can form a kind of tail that pushes them through the host cell's cytoplasm.

DIVERSITY OF BACTERIAL SPECIES

Classification seeks to describe the diversity of bacterial species by naming and grouping organisms based on similarities. Bacteria can be classified on the basis of cell structure, cellular metabolism or on differences in cell components such as DNA, fatty acids, pigments, antigens and quinones. While these schemes allowed the identification and classification of bacterial strains, it was unclear whether these differences represented variation between distinct species or between strains of the same species. This uncertainty was due to the lack of distinctive structures in most bacteria, as well as lateral gene transfer between unrelated species.

Due to lateral gene transfer, some closely related bacteria can have very different morphologies and metabolisms. To overcome this uncertainty, modern bacterial classification emphasises molecular systematics, using genetic techniques such as guanine cytosine ratio determination, genome-genome hybridisation, as well as sequencing genes that have not undergone extensive lateral gene transfer, such as the rRNA gene. Classification of bacteria is determined by publication in the International Journal of Systematic Bacteriology, and Bergey's Manual of Systematic Bacteriology.

The term "bacteria" was traditionally applied to all microscopic, single-celled prokaryotes. However, molecular systematics showed prokaryotic life to consist of two separate domains, originally called Eubacteria and Archaebacteria, but now called Bacteria and Archaea that evolved independently from an ancient common ancestor. These two domains, along with Eukarya, are the basis of the three-domain system, which is currently the most widely used classification system in bacteriology. However, due to the relatively recent introduction of molecular systematics and the analysis of genome sequences, bacterial classification remains a changing and expanding field.

Identification of bacteria in the laboratory is particularly relevant in medicine, where the correct treatment is determined by the bacterial species causing an infection. Consequently, the need to identify human pathogens was a major impetus for the development of techniques to identify bacteria.

The Gram stain, developed in 1884 by Hans Christian Gram, characterises bacteria based on the structural characteristics of their cell walls. The thick layers of peptidoglycan in the "Gram-positive" cell wall stain purple, while the thin "Gram-negative" cell wall appears pink. By combining morphology and Gram-staining, most bacteria can be classified as belonging to one of four groups. Some organisms are best identified by stains other than the Gram stain, particularly mycobacteria or Nocardia, which show acid-fastness on Ziehl-Neelsen or similar stains. Other organisms may need to be identified by their growth in special media, or by other techniques, such as serology.

Culture techniques are designed to promote the growth and identify particular bacteria, while restricting the growth of the other bacteria in the sample. Often these techniques are designed for specific specimens; for example, a sputum sample will be treated to identify organisms that cause pneumonia, while stool specimens are cultured on selective media to identify organisms that cause diarrhoea, while preventing growth of non-pathogenic bacteria. Specimens that are normally sterile, such as blood, urine or spinal fluid, are cultured under conditions designed to grow all possible organisms. Once a pathogenic organism has been isolated, it can be further characterised by its morphology, growth patterns such as and staining.

As with bacterial classification, identification of bacteria is increasingly using molecular methods. Diagnostics using such DNA-based tools, such as polymerase chain reaction, are increasingly popular due to their specificity and speed, compared to culture-based methods. These methods also allow the detection and identification of "viable but nonculturable" cells that are metabolically active but non-dividing. However, even using these improved methods, the total number of bacterial species is not known and cannot even be estimated with any certainty. Attempts to quantify bacterial diversity have ranged from 107 to 109 total species, but even these diverse estimates may be out by many orders of magnitude.

The large number of microorganisms present in the bacteria display a wide diversity of overlapping abilities and these traits do not match up neatly into any form of classification. For example, the metabolic capabilities of the proteobacteria seem almost randomly dispersed throughout the group and do not fall into any discernable arrangement. Therefore, any scheme to organise bacteria will have its problems, and unavoidably, certain abilities appear several times in different places. Sulfate reduction and photosynthesis are two examples of abilities dispersed throughout this domain.

Alpha-Proteobacteria

Proteobacteria are evolutionarily related to each other, but have so evolved that one of the few things they have in common is their gram-negative cell wall. They are one of the largest collections of cultured microorganisms and many microbes important to humans fall into this group. There is tremendous diversity within the group. Strict aerobes, facultative anaerobes and strict anaerobes are found in this family and a number of different cellular morphologies are also present, although they tend to be rod-shaped.

There are five subgroups of Proteobacteria a, ß, g, d, and e and it will examine a few species from each subgroup to give you a sense of the wide variety of microorganisms and metabolisms.

Alpha Subgroup

Rhizobia were originally identified as bacteria that form nodules on leguminous plants and were all lumped into one genus, Rhizobium. However, you wont be surprised to learn that phylogeny based on such a quality happens to be incorrect and phylogenetic evidence has since made it clear this group of microbes is quite diverse. The rhizobia have been split into six different genera based upon 16S rRNA analysis, Rhizobium, Sinorhizobium, Mesorhizobium, Bradyrhizobium, Azorhizobium and Allorhizobium. These plant-nodulating bacteria do not form a self-contained group, but are intermingled with a number of species that do not form symbiotic relationships. In other words, members of the Rhizobium and Bradyrhizobium are less related to each other than to genera that fail to nodulate plants.

Rhizobia are gram-negative chemoheterotrophic organotrophs and are capable of growth on a large array of carbohydrates. They are normally rod-shaped and do not form endospores. They are motile by one polar or subpolar flagellum or by 26 peritrichous flagella, and grow by aerobic respiration of various carbon compounds. They are easily cultivated on medium containing carbohydrates and a considerable portion of this carbohydrate is converted to extracellular slime, which presumably serves a role in the environment.

The behavior of rhizobia toward carbohydrates no doubt reflects the fact that this material is available in the vicinity of plants where rhizobia are often found. Because of photosynthesis, plants produce an excess of carbohydrates that can be available to bacteria that are either in the vicinity of plant roots or have formed a symbiotic relationship in nodules.

Rhizobia can be divided into fast-growing strains, with generation times of less than 6 hours, and slow-growing strains. Fast growing strains utilise the Entner-Meyerhoff-Parnas pathway, while most of the slow-growing rhizobia utilise the Enter-Douderoff pathway. Fast-growing strains are capable of catabolising glucose, galactose, fructose, xylose, mannitol, sucrose, trehalose, and dulcitol. Slow-growing strains can utilise these sugars, except for surcrose, trehalose and dulcitol. Slow-growing strains also tend to have growth factor requirements in addition to the carbon and energy source.

The fast-growing rhizobia have the genes for nodulation and nitrogen fixation on very large plasmids. These plasmids are actually larger than the entire chromosomes of other bacteria and are so large (~1 Mbp) that they were not detected as plasmids for many years until plasmid-handling methods became more sophisticated. This raises the following interesting point. It seems plausible that nodulation was present in the progenitor to rhizobia and its non-nodulating relatives and had been selectively lost over time. It is also possible that nodulation transferred into strains by way of this plasmid.

Certainly the presence of the genes on plasmids, rather than on the bacterial chromosome, makes it easier to understand how they might be lost. However, the plasmid-location suggests another alternative: perhaps the progenitor did not have this ability, but it evolved later on a plasmid. The ability of such plasmids to move among strains might also explain the presence in some but not all members of the larger family. But if this were the case, wouldnt you expect the plasmid to be found in relatively unrelated bacteria as well? Perhaps, but perhaps not.

Possibly, there are other metabolic features in this bacterial family that make them particularly good at metabolising plants products, so that they are the only microbes that can use the information on the plasmid efficiently and are therefore the only ones to retain it. At least this story shows the complexity in trying to make such progenitor identification arguments. The slow-growing strains also have the genes for nitrogen fixation, but in this case they are in their chromosomes. It is possible that these microbes recombined the aforementioned plasmid into their chromosome and it has been retained there.

Rhizobia are called plant symbionts, rather than pathogens, because when they nodulate plants, they actually provide a benefit. They fix nitrogen and supply it to the plant, which provides them with carbon and energy in return. They are capable of forming relationships with a wide variety of legumes such as alfalfa, soybeans and peas. In nitrogen-poor soils, rhizobia give a competitive advantage to their hosts, allowing them to grow in nitrogen poor soil.

There has been substantial scientific interest in the rhizobia since many of the plants they nodulate are important crops around the world and the rhizobia contribute significantly to the overall fertility of the soil. A few years after rhizobia were first described and cultured in 1895, they were used for inoculating legumes by coating the seeds with culture or mixing soil, seeds and culture together. This improved the rate of nodulation of the plants and had a marked effect on their growth. However, the issue has become somewhat complicated as there are now clearly many fields where there is a substantial population of rhizobia in the soil that out-compete those added with the seeds.

In such cases, it is the indigenous bacteria that form nodules and many of these are not particularly effective at fixing nitrogen. Much of the current effort in the field is now directed toward making productive nitrogenfixing rhizobia more competitive in infecting plants. The other approach to improving the agricultural value of biological nitrogen fixation involves nitrogen-fixers other than the rhizobia. These so-called associative nitrogen-fixing bacteria do not form the involved symbiosis with plants and therefore might be more amendable to forming productive relationships with a variety of plants.

Most bacteria divide by binary fission, but prosthecate bacteria reproduce asymmetrically. Their name refers to the prostheca, which is a long appendage found on one of the two cell types. Caulobacter species are gram-negative, non-spore-forming rods. Phylogenetically the prosthecate bacteria are widely divergent, but are more closely related to each other than to non-prosthecate microorganisms.

Bacteria and Other Organisms

Despite their apparent simplicity, bacteria can form complex associations with other organisms. These symbiotic associations can be divided into parasitism, mutualism and commensalism. Due to their small size, commensal bacteria are ubiquitous and grow on animals and plants exactly as they will grow on any other surface. However, their growth can be increased by warmth and sweat, and large populations of these organisms in humans are the cause of body odor.

Certain bacteria form close spatial associations that are essential for their survival. One such mutualistic association, called interspecies hydrogen transfer, occurs between clusters of anaerobic bacteria that consume organic acids such as butyric acid or propionic acid and produce hydrogen, and methanogenic Archaea that consume hydrogen. The bacteria in this association are unable to consume the organic acids as this reaction produces hydrogen that accumulates in their surroundings. Only the intimate association with the hydrogen-consuming Archaea keeps the hydrogen concentration low enough to allow the bacteria to grow.

In soil, microorganisms which reside in the rhizosphere (a zone that includes the root surface and the soil that adheres to the root after gentle shaking) carry out nitrogen fixation, converting nitrogen gas to nitrogenous compounds. This serves to provide an easily absorbable form of nitrogen for many plants, which cannot fix nitrogen themselves. Many other bacteria are found as symbionts in humans and other organisms. For example, the presence of over 1,000 bacterial species in the normal human gut flora of the intestines can contribute to gut immunity, synthesise vitamins such as folic acid, vitamin K and biotin, convert milk protein to lactic acid, as well as fermenting complex undigestible carbohydrates. The presence of this gut flora also inhibits the growth of potentially pathogenic bacteria and these beneficial bacteria are consequently sold as probiotic dietary supplements.

If bacteria form a parasitic association with other organisms, they are classed as pathogens. Pathogenic bacteria are a major cause of human death and disease and cause infections such as tetanus, typhoid fever, diphtheria, syphilis, cholera, food-borne illness, leprosy and tuberculosis. A pathogenic cause for a known medical disease may only be discovered many years after, as was the case with Helicobacter pylori and peptic ulcer disease. Bacterial diseases are also important in agriculture, with bacteria causing leaf spot, fireblight and wilts in plants, as well as Johne's disease, mastitis, salmonella and anthrax in farm animals.

Each species of pathogen has a characteristic spectrum of interactions with its human hosts. Some organisms, such as Staphylococcus or Streptococcus, can cause skin infections, pneumonia, meningitis and even overwhelming sepsis, a systemic inflammatory response producing shock, massive vasodilation and death. Yet these organisms are also part of the normal human flora and usually exist on the skin or in the nose without causing any disease at all. Other organisms invariably cause disease in humans, such as the Rickettsia, which are obligate intracellular parasites able to grow and reproduce only within the cells of other organisms. One species of Rickettsia causes typhus, while another causes Rocky Mountain spotted fever.

Chlamydia, another phylum of obligate intracellular parasites, contains species that can cause pneumonia, or urinary tract infection and may be involved in coronary heart disease. Finally, some species such as Pseudomonas aeruginosa, Burkholderia cenocepacia, and Mycobacterium avium are opportunistic pathogens and cause disease mainly in people suffering from immunosuppression or cystic fibrosis.

Bacterial infections may be treated with antibiotics, which are classified as bacteriocidal if they kill bacteria, or bacteriostatic if they just prevent bacterial growth. There are many types of antibiotics and each class inhibits a process that is different in the pathogen from that found in the host. An example of how antibiotics produce selective toxicity are chloramphenicol and puromycin, which inhibit the bacterial ribosome, but not the structurally different eukaryotic ribosome.

Antibiotics are used both in treating human disease and in intensive farming to promote animal growth, where they may be contributing to the rapid development of antibiotic resistance in bacterial populations. Infections çan be prevented by antiseptic measures such as sterilisating the skin prior to piercing it with the needle of a syringe, and by proper care of indwelling catheters. Surgical and dental instruments are also sterilised to prevent contamination and infection by bacteria. Disinfectants such as bleach are used to kill bacteria or other pathogens on surfaces to prevent contamination and further reduce the risk of infection.

Archaea

The Archaea, or archaebacteria, are a major group of microorganisms. Like bacteria, archaea are single-celled organisms that lack nuclei and are therefore prokaryotes, classified in kingdom Monera in the traditional five-kingdom taxonomy. Although there is still uncertainty in the phylogeny, Archaea, Eukaryota and Bacteria are the fundamental classifications of what is called the three-domain system.

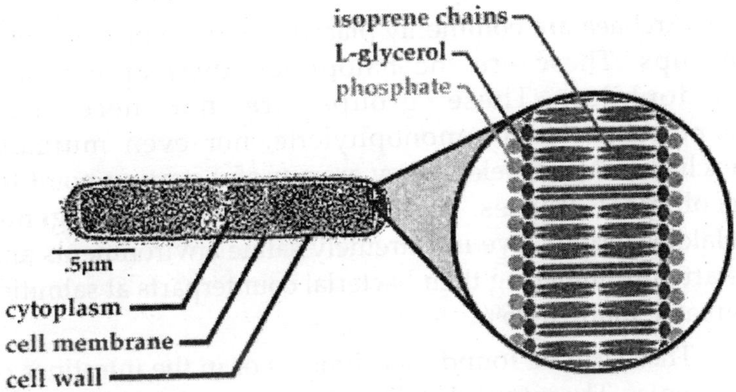

Basic Archaeal Structure

Although their prokaryotic features are diagnostic of that clade, archaea are more closely related to eukaryotes than to bacteria. To account for this, archaeans and eukaryotes are grouped together in the clade Neomura, which is thought to have arisen from gram-positive bacteria. Archaea were originally described in extreme environments, but have since been found in all habitats and may contribute up to 20% of total biomass. A single individual or species from this domain is called an archaeon, while the adjectival form is archaeal or archaean.

Multiple archaeans are extremophiles, and some would say this is their ecological niche. They can survive high temperatures, often above 100°C, as found in geysers, black smokers, and oil wells. Some are found in very cold habitats and others in highly saline, acidic, or alkaline water. Mesophiles favour milder conditions in marshland, sewage and soil. Many methanogenic archaea are found in the digestive tracts of animals such as ruminants, termites, and humans. As of 2007, no clear examples of archaeal pathogens are known, although a relationship has been proposed between the presence of some methanogens and human periodontal disease.

Archaea are commonly placed into three physiological groups. These are the halophiles, thermophiles and acidophiles. These groups are not necessarily comprehensive or monophyletic, nor even mutually exclusive. Nonetheless, they are a useful starting point for ecological studies. Halophiles, including the genus Halobacterium, live in extremely saline environments and start outnumbering their bacterial counterparts at salinities greater than 20-25%.

These can be found in sediments or in the intestines of animals. Thermophiles live in places that have high temperatures, such as hot springs. Where optimal growth occurs at greater than 80°C, the archaeon is a hyperthermophyle, and the highest recorded temperature survived was 121°C. Although thermophilic bacteria predominate at some high temperatures, archaea generally have the edge when acidity exceeds pH 5. True acidophiles withstand pH 0 and below.

Recently, several studies have shown that archaea exist not only in mesophilic and thermophilic environments but are also present, sometimes in high numbers, at low temperatures as well. It is increasingly becoming recognised that methanogens are commonly present in low-temperature environments such as cold sediments. Some studies have even suggested that at these temperatures the pathway by which methanogenesis occurs may change due to the thermodynamic constraints imposed by low temperatures.

Perhaps even more significant are the large numbers of archaea found throughout most of the world's oceans, a predominantly cold environment. These archaea, which belong to several deeply branching lineages unrelated to those previously known, can be present in extremely high numbers (up to 40% of the microbial biomass) although almost none have been isolated in pure culture. Currently we have almost no information regarding the physiology of these organisms, meaning that their effects on global biogeochemical cycles remain unknown. One recent study has shown, however, that one group of marine crenarchaeota are capable of nitrification, a trait previously unknown among the archaea.

Morphology and Physiology

Individual archaeans range from 0.1 μm to over 15 μm in diameter, and some form aggregates or filaments up to 200 μm in length. They occur in various shapes, such as spherical, rod-shape, spiral, lobed, or rectangular. Archaea have no murein in their cell walls. Recently, a species of flat, square archaean that lives in hypersaline pools has been discovered.

Archaea are similar to other prokaryotes in most aspects of cell structure and metabolism. However, their genetic transcription and translation—the two central processes in molecular biology —do not show many typical bacterial features, and are in many aspects similar to those of eukaryotes. For instance, archaeal translation uses eukaryotic-like initiation and elongation factors, and their transcription involves TATA Binding Proteins and TFIIB as in eukaryotes. Many archaeal tRNA and rRNA genes harbor unique archaeal introns which are neither like eukaryotic introns, nor like bacterial introns.

Several other characteristics also set the Archaea apart. Like bacteria and eukaryotes, archaea possess glycerol-based phospholipids. However, three features of the archaeal lipids are unusual:

— The archaeal lipids are unique because the stereochemistry of the glycerol is the reverse of that found in bacteria and eukaryotes. This is strong evidence for a different biosynthetic pathway.

— Most bacteria and eukaryotes have membranes composed mainly of glycerol-ester lipids, whereas archaea have membranes composed of glycerol-ether lipids. Even when bacteria have ether-linked lipids, the stereochemistry of the glycerol is the bacterial form. These differences may be an adaptation on the part of Archaea to hyperthermophily. However, it is worth noting that even mesophilic archaea have ether-linked lipids.

— Archaeal lipids are based upon the isoprenoid sidechain. This is a five-carbon unit that is also common in rubber and as a component of some bacterial and eukaryotic vitamins. However, only the archaea incorporate these compounds into their cellular lipids, frequently as C-20 (four monomers) or C-40 (eight monomers) side-chains. In some archaea, the C-40 isoprenoid side-chain is long enough to span the membrane, forming a monolayer for a cell membrane with glycerol phosphate moieties on both ends. Although dramatic, this adaptation is most common in the extremely thermophilic archaea.

Although not unique, archaeal cell walls are also unusual. For instance, in most archaea they are formed by surface-layer proteins or an S-layer. S-layers are common in bacteria, where they serve as the sole cell-wall component in some organisms (like the Planctomyces) or an outer layer in many organisms with peptidoglycan. With the exception of one group of methanogens, archaea lack a peptidoglycan wall.

Archaeans also have flagella that are notably different in composition and development from the superficially similar flagella of bacteria. The bacterial flagellum is a modified type III secretion system, while archeal flagella resemble type IV pilli which use a sec dependent secretion system somewhat similar to but different from type II secretion system.

Metabolism

Archaea exhibit a variety of different types of metabolism; there are nitrifiers, methanogens and anaerobic methane oxidisers. Methanogens live in anaerobic environments and produce methane. Of note are the halobacteria, which use light to produce energy. Although no archaea conduct photosynthesis with an electron transport chain; light-activated ion pumps like bacteriorhodopsin and halorhodopsin play a role in generating ion gradients, which are harnessed into adenosine triphosphate (ATP).

Genetics and Propagation

Archaea have one circular chromosome although up to 30% of their genetic material may be contained in plasmids, as evidenced by comparisons of GC content. Archaea can reproduce by binary and multiple fission, fragmentation, and budding.

Phylogeny

Archaea are divided into two main groups based on rRNA trees, the Euryarchaeota and Crenarchaeota. Three other groups have been tentatively created for certain environmental samples, the peculiar species Nanoarchaeum equitans, discovered in 2002 by Karl Stetter, and the Archael Richmond Mine Acidophilic Nanoorganisms groups discovered by Brett Baker, but their affinities are uncertain.

Woese argued that the bacteria, archaea, and eukaryotes each represent a primary line of descent that diverged early on from an ancestral progenote with poorly developed genetic machinery. Later he treated these groups formally as domains, each comprising several kingdoms. This division has become very popular, although the idea of the progenote itself is not generally supported. Some biologists, however, have argued that the archaebacteria and eukaryotes arose from specialised eubacteria.

The relationship between Archaea and Eukarya remains an important problem. Aside from the similarities noted above, many genetic trees group the two together. Some place eukaryotes closer to Euryarchaeota than Crenarchaeota are, although the membrane chemistry suggests otherwise. However, the discovery of archaean-like genes in certain bacteria, such as Thermotoga, makes their relationship difficult to determine, as horizontal gene transfer may have occurred.

The Archaea should not be confused with the geological term Archean eon, also known as the Archeozoic era. This refers to the primordial period of earth history when Archaea and Bacteria were the only cellular organisms living on the planet. Probable fossils of these microbes have been dated to almost 3.5 billion years ago, and the remains of lipids that may be either archaean or eukaryotic have been detected in shales dating from 2.7 billion years ago.

The last common ancestor of Bacteria and Archaea was probably a non-methanogenic thermophile, raising the possibility that lower temperatures are extreme environments in archaeal terms, and organisms that can survive in cooler environments evolved later on.

Protists

Animals, plants, fungi, and protists are eukaryotes, organisms whose cells are organised into complex structures by internal membranes and a cytoskeleton. The most characteristic membrane bound structure is the nucleus. This feature gives them their name, also spelled "eucaryote", which comes from the Greek e?, meaning good/true, and meaning nut, referring to the nucleus. In the nucleus the genetic material, DNA, is arranged in chromosomes. Many eukaryotic cells also contain membrane-bound organelles such as mitochondria, chloroplasts and Golgi bodies. Eukaryotes often have unique flagella made of microtubules in a 9+2 arrangement.

Cell division in eukaryotes is also different from organisms without a nucleus. This process involves separating the duplicated chromosomes, through movements directed by microtubules. There are two types of these division prcesses. In mitosis one cell divides to produce two genetically identical cells.

In meiosis, which is required in sexual reproduction, one diploid cell undergoes a process of recombination between each pair of parental chromosomes, and then two stages of cell division, resulting in four haploid cells each of which has only a single complement of chromosomes, each one being a unique mix and match of the corresponding pair of parental chromosomes.

Eukaryotes appear to be monophyletic and thus make up one of the three domains of life. The two other domains: bacteria and archaea share none of the above features, though the eukaryotes do share some aspects of their biochemistry with the archaea, and as such, are grouped with the archaea in the clade Neomura.

There are many different types of eukaryotic cells, though animals and plants are the most familiar eukaryotes, and thus provide an excellent starting point for understanding eukaryotic structure. Fungi and many protists have some substantial differences, however.

Animal Cell

An animal cell is a form of eukaryotic cell which make up many tissues in animals. The animal cell is distinct from other eukaryotes, most notably plant cells, as they lack cell walls and chloroplasts, and they have smaller vacuoles. Due to the lack of a rigid cell wall, animal cells can adopt a variety of shapes and a phagocytic cell can even engulf other structures.

Plant cells are quite different from the cells of the other eukaryotic organisms. Their distinctive features are:

— A large central vacuole (enclosed by a membrane, the tonoplast), which maintains the cell's turgor and controls movement of molecules between the cytosol and sap.
— A cell wall made up of cellulose and protein, and in many cases lignin, and deposited by the protoplast on the outside of the cell membrane. This contrasts with the cell walls of fungi, which are made of chitin, and prokaryotes, which are made of peptidoglycan.
— The plasmodesmata, linking pores in the cell wall that allow each plant cell to communicate with other adjacent cells. This is different from the network of hyphae used by fungi.
— Plastids, especially chloroplasts that contain chlorophyll, the pigment that gives plants their green color and allows them to perform photosynthesis.
— Plant groups without flagella also lack centrioles that are present in animal cells.

Fungal cells are most similar to animal cells, with the following exceptions.

— A cell wall made of chitin.
— Less definition between cells. Higher fungal cells have porous separations called septa which allow the passage of cytoplasm, organelles, and sometimes, nuclei. Primitive fungi have no such divisions, and each organism is essentially a giant supercell. These fungi are described as coenocytic.
— Only the most primitive fungi, chytrids, have flagella.

REFERENCES

Brack, A. (Ed.), *The Molecular Origins of Life: Assembling Pieces of the Puzzle, Cambridge University Press, London,* 1998

Brock, T. D., *Milestones in Microbiology*, Prentice-Hall, Englewood Cliffs, NJ, 1961.

Pelczar, Michael J., *Microbiology*, New York: McGraw-Hill, 1958.

3

Microbial Genetics

Microorganisms have the ability to acquire genes and thereby undergo the process of recombination. In recombination, a new chromosome with a genotype different from that of the parent results from the combination of genetic material from two organisms. This new arrangement of genes is usually accompanied by new chemical or physical properties.

In microorganisms, several kinds of recombination are known to occur. The most common form is general recombination, which usually involves a reciprocal exchange of DNA between a pair of DNA sequences. It occurs anywhere on the microbial chromosome and is typified by the exchanges occurring in bacterial transformation, bacterial recombination, and bacterial transduction.

A second type of recombination, called site-specific recombination, involves the integration of a viral genome into the bacterial chromosome. A third type is replicative recombination, which is due to the movement of genetic elements as they switch position from one place on the chromosome to another.

The principles of recombination apply to prokaryotic microorganisms but not to eukaryotic microorganisms. Eukaryotes exhibit a complete sexual life cycle, including meiosis. In this process, new combinations of a particular gene form during the process of crossing over. This process occurs between homologous chromosomes and is not seen in bacteria, where only a single chromosome exists. Much of the work in microbial genetics has been performed with bacteria, and the unique features of microbial genetics are usually those associated with prokaryotes such as bacteria.

NATURAL SELECTION

Over the years we have progressed with genetic dissection of the diverse morphs. The evidence (which ranges from estimates of mutation rate, to identification of many causal mutations) points to diversification being wrought by genetic adaptation: natural selection acting on spontaneous mutation.

Upon entering a novel environment replete with ecological opportunity, the ancestral genotype experiences strong diversifying selection (which is not experienced in the spatially unstructured environments). Selection favours particular mutants because of the fitness benefits they engender. Competition among competing beneficial mutations—particularly for oxygen—fuels the diversification process.

A fortuitous feature of the Pseudomonas populations is the correspondence between niche specialist genotype and colony morphology on agar plates. This means that it has been possible to look at the dynamics of genetic diversification simply by scoring the frequencies of colony variants. Put another way, it has been impossible to ignore evolution. Had evolutionary diversification not manifested in a visible way, we would have overlooked the phenomenon. This leads us to wonder whether evolution in bacterial populations is overlooked, moreover, whether it being overlooked is problematic.

With certainty, evolution occurs, is overlooked, ignored, or incorrectly interpreted. Depending on the particular questions asked, this may or may not matter. The way forward is first to recognise that no environment is benign. For a freshly isolated bacterium entering the laboratory environment selection for mutants that grow more rapidly in vitro will be intense. Moreover, initial fitness improvements are likely to involve loss of traits not required in vitro, the problem here is that many of these traits may be relevant to the life of the bacterium in its natural environment.

Changes of Culture

In many circumstances there may be little that can be done other than to recognise the fact that evolution occurs and to account for its effects in experimental design and interpretation of results. However, in some situations failure to recognise genetic adaptation and its effects can be seriously problematic, resulting in the formulation of implausible hypotheses, the design and execution of meaningless mutant hunts and incorrect conclusions about the ecological significance of genes and regulatory networks—even of the organism itself.

While no area of microbiology is immune from this difficulty, one area that seems to be at risk of being misled (if it hasn't already succumbed) is that concerned with 'biofilms'. Studies in this area rely heavily on experiments performed in flow chambers—environments that impose intense selection for surface-colonising mutants: a cell that doesn't stick is washed from the chamber and is effectively dead. Moreover, flow chamber experiments are typically run over the course of days, thus providing ample opportunity for natural selection to operate. At the very least flow chamber experiments should include a control in which the derived population (the population taken from the flow chamber at the end of the experiment) is competed directly with the ancestral population (the population used to found the original flow chamber).

Two processes are relevant to how we think about the response of bacterial populations to environmental change: phenotypic acclimation and genetic adaptation by natural selection (in fact the two processes are not mutually exclusive). For reasons that are complex and worthy of further exploration in their own right, microbiologists have tended to overlook or misinterpret genetic adaptation by natural selection, assuming that phenotypic acclimation is sufficient to account for the majority of responses to environmental challenge. This oversight has serious implications for how we think about and study microbes: for the kinds of models we develop to inform and instruct the investigations, for the way we think about evolution in bacteria and for the understanding of the function of bacteria in natural environments.

RNA VIRUSES

With RNA viruses, we have the opportunity to observe evolution in action, and empirically test those principles of mutation, inheritance and fitness selection on which Darwin's theory depends. Three attributes make RNA viruses particularly suitable objects of study.

— First, they replicate extremely rapidly, with generation times of as little as 30 minutes in the case of bacteriophages.

— Second, their extremely error-prone replication provides a rich source of mutations on which selection acts.

— Third, they can achieve huge population sizes (up to 1012 replicating virions in the body of an HIV or HCV-infected individual), leading to intense fitness selection far beyond that found in the evolution of larger organisms such as mammals.

In both natural history and in vitro studies, there are several dramatic examples of 'positive' or 'Darwinian' selection, in which selection for, and inheritance of, phenotypic changes in mutant virus leads to complete population replacements over short observation periods. For example, each time an HIV-infected individual is treated with the antiviral agent AZT, a clinically resistant population of HIV rapidly emerges bearing the same 2-4 amino acid changes in the part of the reverse transcriptase enzyme that determines substrate recognition.

Similarly, many virus infections show evidence for positive selection in response to activity of the immune system. Over the course of an epidemic, the external haemagglutinin of influenza A shows dramatically accelerated evolution of infection, keeping it ahead of the neutralizing antibody response mounted by the host. Primary infection with HIV is associated with a burst of rapid sequence change that eliminates specific MHC class I recognition motifs in several virally encoded proteins, therefore preventing destruction of the infected cell by cytotoxic T cells.

Genetic Variation

Surprisingly, sequence changes occurring in response to phenotypic selection makes very little contribution to the genetic diversity of viruses and other organisms, measured at the level of DNA or RNA sequences. Indeed, the majority of sequence change that occurs over time has no significant effect on phenotype (i.e. it is 'neutral') and becomes fixed in the population by chance. In what seems a baffling contrast to the dramatic phenotypic changes made by single mutations, the process of drift over time leads to genotypes or species of RNA viruses differing from each other by 30-40 % or even more in nucleotide sequences, each remaining remarkably similar in their biological properties.

For example, the coding sequence in the six genotypes of HCV shows greater than 30 % divergence from each other, but all HCV variants have retained identical genome organisations and replication cycles, and each retains HCV's peculiar ability to persist in humans and remains largely unchanged in their capacity to cause slowly progressive liver disease.

Although the fixation of phenotypically neutral substitutions that takes place on sequence drift should occur at a constant rate over time, it is nevertheless difficult to date times of divergence of different virus variants, genotypes or species. From the epidemiological distribution of different genotypes of GB virus-C (GBV-C), a non-pathogenic flavivirus cousin of HCV, it appears that its current sequence diversity (11-13 %) arose extremely slowly, potentially as long as the 150,000 year period during which anatomically modern human population groups re-populated the world from Africa.

RNA Structure

In the case of GBV-C, HCV and other persistent RNA viruses, we have recently found that their sequence drift is limited by a lack of truly 'neutral' sites.

The nucleotide sequences of their genomes are organised to form evolutionarily conserved, extensive RNA secondary structures, termed genome-scale organised RNA structures (GORS), which may in some way modulate the activity of host-cell defences that recognise doublestranded RNA. GORS is destroyed by even minimal nucleotide changes that would typically be regarded as 'neutral', such as those at third codon positions that do not alter the encoded protein sequence. GORS may therefore place some viruses in an evolutionary 'straightjacket' in which only a minority of changes that retain base pairing are tolerated. Such RNA viruses are clearly not the progressively evolving and ephemeral entities of popular imagination.

EVOLUTION OF ANTIFUNGAL RESISTANCE IN CANDIDASPECIES

The problems of antifungal resistance are different in many

ways from those of antibacterial resistance. Fungi differ from bacteria in one very important respect: they have, so far as we know, no plasmids or other natural mechanism capable of transferring genetic material between strains. This means that a strain of C. albicans that has become resistant to an antifungal agent cannot pass on the gene(s) encoding the resistant phenotype to other strains.

The potential for rapid emergence of resistance that has been seen with many types of bacteria is therefore much smaller in C. albicans. We have already seen how antifungal resistance emerges in clinical practice. Before effective therapy to combat AIDS was available, around 20 % of patients infected with HIV who were repeatedly treated with the agent fluconazole for Candida infections of their mouths ended up with infecting strains that were many-fold less susceptible to fluconazole than the strains that caused the first bouts of infection.

Yet it was rare to find any replacement of an individual's infecting strain with another that clearly differed in molecular typing tests: resistance almost always emerged by a process of 'micro-evolution' within each patient's own isolate responding locally to the selective pressure of fluconazole exposure. In the best known example of this process, a set of 17 consecutive oral isolates of C. albicans was obtained over a 2-year period from one patient as his $CD4^+$ cell count deteriorated.

The fluconazole susceptibility of each new isolate was as low as, or lower than its predecessor. Yet the isolates were indistinguishable by a variety of DNA typing tests. The mechanisms of resistance expressed in each of the isolates were determined. In isolate number 3, the gene MDR1, a fluconazole efflux pump, was overexpressed. In isolate number 13, mutations were detected in the gene ERG11, which encodes the enzyme that is the target for fluconazole; the mutation was found in both alleles of the gene (C. albicans is diploid).

Moreover, the mutated ERG11 was also over-expressed. Isolates numbers 16 and 17 added over-expression of another gene encoding an efflux pump, CDR1, to the other mechanisms of resistance. So these last isolates had evolved to express at least four different mechanisms of fluconazole resistance in each cell. However, there is no evidence that these multiple-resistant strains were transferred to other AIDS patients. Of course, patient to patient transfer of resistant strains can occur, but this means of spread is far less efficient than would be the case if the fungi could transfer DNA from strain to strain.

Loss of Fitness

Evolution to an antifungal resistant genotype is not a cost-free experience for the fungus. Fitness, typically defined as reproductive output over a specific time interval, may be reduced in a strain that has evolved to antifungal resistance as compared with its susceptible predecessor. Trade-offs of fitness in exchange for acquisition of survival-enhancing characters in a specific environment form an inevitable part of evolutionary processes. With C. albicans, Dr James Anderson and his colleagues in Canada have experimentally investigated the evolution of fitness and susceptibility in initially clonal populations of a fluconazole- sensitive isolate that was sequentially subcultured in a medium with and without additions of fluconazole.

They found that fluconazole exposure led to different levels of resistance development in different subpopulations, that resistant populations initially paid a penalty in fitness for the mutation to resistance, but that fitness returned to normal with further exposure to the fluconazole-containing environment. Microarray experiments with the sensitive and resistant strains showed that expression of no fewer than 301 genes altered significantly as the populations adapted to fluconazole resistance! The expression changes could be assigned to three fairly general patterns, which were associated with different stages of resistance development.

Over-arching Mechanisms

Although C. albicans has a diploid genome, recent research has shown that the fungus contains genes homologous to the mating-type genes of Saccharomyces cerevisiae, and several groups of investigators have now succeeded in inducing isolates that are homozygous a or átype at the mating-type-like (MTL) locus to undergo a mating process, though not yet with the completion of meiosis. Intriguingly, clinical isolates of C. albicans that are fluconazole-resistant are far more likely than others to be homozygous at the MTL.

As DNA-level strain typing methods for C. albicans have progressed, ever more reliable population structure analyses of the species show that the majority of isolates of the yeast can be assigned to one of four major, closely related clades. Dr David Soll and his group in Iowa have shown that almost all isolates of C. albicans expressing resistance to flucytosine belong to a single clade. Moreover, all flucytosine-resistant isolates from that clade owe their resistance to a single point mutation in a target gene. Perhaps resistance mechanisms to other agents will ultimately prove to be restricted to particular clades.

Antifungal Resistance

Studies of the evolution of antifungal resistance in C. albicans have reached a particularly exciting point. We know that mutation to one or more triazole resistance mechanisms occurs readily in response to fluconazole exposure, and that the changes involved can require collateral alterations in levels of expression in as much as 5% of the C. albicans genome.

And we know that a common mechanism of mutation to flucytosine resistance pertains to isolates in one of the four major clades of the global C. albicans population. What we do not know is how these processes are triggered and regulated; why strains need to mutate to constitutively over-express efflux pumps (why don't they just over-express when the drug is present and turn down the gene expression when it is not?); why just one clade of related strains should have a near monopoly on mutation to flucytosine resistance?

Research on C. albicans has been well supported and has itself evolved rapidly over the past 20 years. We may soon have answers to some of the intriguing questions posed by the evolution of antifungal resistance in Candida species; meanwhile we can take comfort in recent surveillance data which show unequivocally that antifungal resistance among clinical isolates of Candida species is growing slowly, if at all. We appear to have time to understand the mechanisms well before any clinical resistance problem arises.

SERIAL ENDOSYMBIOTIC THEORY (SET) AND COMPOSITE INDIVIDUALITY

Transition from Bacterial to Eukaryotic Genomes

The Russian-American Drosophila geneticist Theodosius Dobzhansky wrote that 'nothing in biology makes sense except in the light of evolution'. As Darwin noted, our classification systems should become genealogies. If our taxa classify, identify and name life accurately, our grouping will reflect evolution; this is possible because strong inferences concerning the past are embodied in the living. The contribution to evolution of microbiology (sensu lato, the study of both bacteria and their protist descendants) has only recently begun to be appreciated. The cells of microbes are the units of life, hence the recognition of their importance in their own evolution and evolution of larger life forms is bound to increase.

The living world unambiguously is divided into two definitive never-overlapping categories: prokaryotes and eukaryotes. In spite of the immensely useful 'three-domain' 16S rRNA classification scheme proposed by Carl Woese, only two fundamentally different kinds of life exist on Earth. No evidence from either the fossil record or the living world can be mustered for any 'progenote' or other deviation from the prokaryote-eukaryote rule.

This prokaryote-eukaryote divide (Prokarya-Eukarya) remains the largest discontinuity in the living world. First recognised by Edouard Chatton and first analysed by the Delft School of Microbiology the list of differences between Archaebacteria, Eubacteria and Eukarya unequivocally shows that the two prokaryote groups are far more closely related to each other than each of them is to any eukaryote. The cell, whether bacterial [where its genome is chromonemal and unbounded by membrane] or mitotic (nucleated, where a lipoprotein membrane bounds the proteinaceous chromosome) is the unit of all life.

No system of matter and energy flow less complex than a cell is alive. The presence of the nucleus is the only feature that uniquely defines the eukaryotes and distinguishes them from bacteria. The origin of the bacterial cell is the origin of life itself, whereas Serial Endosymbiotic Theory (SET) describes the subsequent origin of the nucleated cell by symbiogenesis. To proceed we need to explain how the ecological concept of 'symbiosis' differs from the evolutionary term 'symbiogenesis'. Symbiosis refers to the living together of organisms of different species. Endosymbiosis, a topological condition, is a kind of symbiosis where one partner lives inside of another.

Symbioses usually, if not always, have environmentally contingent outcomes. Symbiosis, not an evolutionary process per se, refers to physiological, temporal or topological associations with environmentally determined fates. Symbiogenesis, however, implies the appearance of new tissues, new organs, physiologies or other new features that result from protracted symbiotic association. Two great classes of eukaryotic cell organelles, plastids and mitochondria, evolved symbiogenetically. Oxygen-respiring, heterotrophic á-proteobacteria were probably phagocytosed by anaerobic motile protists (like today's mastigamoebae).

Genetic and metabolic redundancies were selected against as once free-living eubacteria evolved into the organelles we recognise as mitochondria. The descendants of this merger include most heterotrophic protists such as most amoebae, cryptomonads, chilomonads and chytrids, oomycotes (like Phytopthora infestans, the potato blight organism). No doubt some motile protists ingested, but failed to digest, food—cyanobacterial cells -that eventually became symbionts. The retention of undigested cyanobacteria in well-lit waters led to permanent unions in which, once again, natural selection favoured the reduction of genetic and metabolic redundancy.

In this way algae, eukaryotic organisms that bear photosynthetic organelles in their cytoplasm, evolved and some became, eventually, the ancestors to the land plants. The Apicomplexa (a phylum which Plasmodium, the genus to which the malarial parasite is assigned) apparently evolved from one lineage of such algae. The members of this phylum, including Toxoplasma have retained a residuum plastid with its DNA, but they are no longer capable of photosynthesis. The principle of 'use it or lose it' can be invoked.

Natural selection does not plan ahead; the unused plastids that began as cyanobacteria were severely reduced as they evolved. The striking resemblance of some free-living bacteria (such as cyanobacteria) to certain intracellular organelles (such as green algal chloroplasts) bolsters the concept that certain bacteria have been trapped inside other cells for millennia. With respect to the acquisition of mitochondria from free-living β-proteobacteria and that of plastids from free-living cyanobacteria, no one any longer doubts that the oxygen respiratory and photosynthesising organelles evolved by symbiogenesis.

Modern symbioses, both intra- and extracellular, that can be subjected to experimental analysis are of extraordinary importance for understanding evolution. How cells merge and how redundancy is reduced is especially relevant to the appearance of the first eukaryotes (which, by definition, were the first protists). Ironically, although most disease conditions are variations on the general theme of cyclical symbioses, few protistologists and microbiologists are familiar with the insightful, burgeoning literature that analyses these nearly ubiquitous associations.

The study of genomics and proteomics will confirm or falsify this historical reconstruction that was made primarily based on observations of live organisms. Organismic biology coupled with direct knowledge of the fossil record are indispensable to evolutionary reconstruction. The techniques of molecular biology and sequence analysis by themselves are inadequate to the creation of testable evolutionary hypotheses.

Evolution of the Tethered Nucleus

The Proterozoic Eon (2500-541 million years ago) was the backdrop for the appearance of cells at the protist level of organisation. All eukaryotes, in the SET, are products of symbiogenesis whereas no prokaryote cell evolved by merger of whole-cell predecessors. Sulfur syntrophy, we hypothesise, united thermoplasmic archaebacteria (such as Thermoplasma acidophilum) with motile Spirochaeta-like eubacteria in the evolution of the karyomastigont organellar system of swimming protists.

The first eukaryotes were composed of at least two integrated bacterial genomes with tethered nucleus (nuclear connector or rhizoplast, kinetosome-axoneme). This organellar system called the 'karyomastigont' has been known to protozoologists since it was first described by C. Janicki in the 1930s. Think of it as the nucleus attached by fibres to the kinetosome-centrioles of the undulipodia: what those who know nothing about bacteria would call 'the nucleus attached to basal bodies and their flagella, usually two or four'. We interpret this organellar system [the karyomastigont of the so-called flagellates (which should be called mastigotes), zoospores of water molds and slime molds, many motile algal cells, etc.] as a legacy of that first genomic integration of these bacteria. The evolution of mitosis with its histone-coated, nucleosome-studded chromatin occurred under anoxic, acidic, organic-rich, and probably muddy conditions prior to the symbiotic acquisition of oxygen-respiring á-proteobacteria that became the mitochondria.

The nucleus with the combined genomes of at least two different prokaryotes evolved to assure genetic continuity of the now integrated archae- and eubacterial symbionts. The nucleus itself began in the karyomastigont as the integrated symbionts, in an act homologous to conjugation between very different bacteria, fused to form the first eukaryote. The untethering of the nucleus in many lineages led to the free nuclei.

Free nuclei seen today in animals, plants and fungi we interpret as the derived state. Tethered nuclei evolved simultaneously with the first protist. No missing links need to be hypothesised. Certain amitochondriate eukaryotes always were confined to anoxic environments and never had mitochondria. The nucleus, in this scenario preceded both the mitochondria and the plastids. Indeed, in the bowels of xylophagous insects (wood-ingesting roaches and termites) and in anoxic muds all 'intermediate stages' that we envision as steps in the origin of nucleated cells are still found today.

Comparative Genomics of *E. coli*

E. coli is notorious for two reasons; first, many people have encountered one or more of the nasty strains responsible for gastrointestinal disease or urinary tract infection, and second, tamed to the Petri dish in the 1920s, the K-12 strain became a favourite of geneticists and molecular biologists who discovered and elaborated fundamental genetic and biochemical processes by studying this organism. The realisation that basic cellular activities such as metabolism, DNA replication and gene expression are achieved using homologous systems in organisms as different as bacteria and man led to extensive investigation of their molecular basis in *E.coli,* owing to its easy manipulation in the lab.

In this way, *E. coli,* along with other experimentally tractable organisms such as fruit flies and yeast, became premier 'model organisms'; systems where scientists could extract maximal information about critical processes with minimal experimental effort. Not surprisingly, *E. coli* K-12 was among the first organisms targeted for complete genome sequencing due to its prominent role as a model organism and industrial tool. Genomes of additional strains of *E. coli* have been sequenced because of their impact on human health. Comparisons of these genomes revealed the genetic diversity underlying the phenotypic diversity of this species.

Gene gain and loss is extensive, providing different lineages with distinct metabolic, pathogenic and other capabilities. The variable regions identified through comparative genomics are serving as candidates for further research on the evolution, epidemiology and molecular biology of *E. coli.* As more *E. coli* genome sequences accumulate, we continue to discover new genes in the population. Ongoing genome projects directed at additional distinct pathogenic *E. coli* will provide an even richer data set. Perhaps the future of *E. coli* as a model organism is *E. coli* as a model bacterial species. The *E. coli* K-12 strain MG1655 genome sequence was published in 1997. It marked the culmination of a project that spanned the transition from tedious radioactive methods to automated fluorescent sequencing and other technological innovations, advancing genome sequencing in general and the efficient completion of additional *E. coli* genomes. In 2001, two sequences were published for enterohaemorrhagic *E. coli* strains of the serogroup O157:H7, one isolated from a contaminated ground beef lot associated with a disease outbreak in the US and the other from the single largest reported outbreak of all time, a 1996 incident involving school lunches in Japan. In 2003, genome sequences were published for an *E. coli* strain (CFT073) associated with urinary tract infections and two strains of Shigella flexneri serotype 2a (2457T and 301).

Differences among Genomes

A comprehensive understanding of the commonalities and differences among these genomes has not yet been achieved, in large part because of the massive amount of genomic variability within the species. Even before the first *E. coli* genome was complete there was substantial knowledge of conservation in function, regulation and chromosomal positions of genes involved in core processes like central metabolism, transcription and translation as well as examples of genes responsible for differences between strains, often contained on mobile genetic elements such as plasmids. As larger pieces of the K-12 DNA sequence became available, examination of patterns in DNA sequence composition and codon usage revealed evidence of atypical 'islands' nestled among well studied genes. Despite evidence hinting at the variability within this species most researchers did not expect the magnitude of changes observed when compared with the next *E. coli* genomes.

The two *E. coli* O157:H7 genomes come from very closely related strains and, as expected, their genomes are extremely similar. However, comparison of either O157:

H7 sequence to *E. coli* K-12 reveals that an extraordinary amount of gene loss and gain has occurred since these strains last shared a common ancestor roughly 5 million years ago. The 5.5 million base pair (Mbp) O157:H7 genome is nearly 1 Mbp larger than the 4.6Mbp *E. coli* K-12 sequence. This much was evident in advance of genome sequencing. What was more surprising was while roughly 4.1 Mbp of the chromosome was very similar (~98.5% identical) between either O157:H7 and K-12, this conserved 'backbone' was punctuated with hundreds of 'islands' of sequences specific to one strain or the other.

For example, O157:H7 strain EDL933 has over 1.4 Mbp of DNA without a counterpart in the K-12 genome, and these sequences are found clustered into 177 regions ranging from 50 to nearly 90,000 bp in length. Among more than a thousand genes encoded in these islands are some that were previously associated with O157:H7 virulence and many new candidate pathogenicity factors, such as iron utilisation and hostcell adherence-associated genes, that are now under active investigation. A surprising discovery was two large islands, each containing (nearly identical, apparently functional) genes encoding urease—and this in a strain typically characterised by its lack of urease activity in clinical assays.

A large fraction of the genomic differences can be accounted for by the activity of mobile genetic elements. Nearly 40 % of the O157-specific elements are found in one of at least 18 cryptic prophage or the one intact bacteriophage (933W). Interestingly, some of the most characteristic virulence genes of O157:H7, those for the Shiga-like toxins, are encoded within phages and evidence is mounting for additional phage-associated virulence determinants.

Horizontal Gene Transfer

If we consider the comparison from the *E. coli* K-12 perspective, it is important to note that many of the regions of the K-12 chromosome without counterpart in O157:H7 correspond to those previously noted islands of atypical base composition. It is believed that these odd sequences come from the occasional transfer of DNA from other species and integration into the *E. coli* chromosome, known as horizontal transfer. The strange base composition of these sequences is due to a high degree of variability in this characteristic between bacterial families. It is also significant that many of the K-12 islands are found at the same position in the chromosome as O157:H7-specific islands, and these also have atypical base composition. Taken together, these observations suggest that the elements specific to each strain accumulated over time by repeated horizontal gene transfer, frequently with successive transfers of different elements into the same slot of the core chromosome, descended from the common ancestor of all *E. coli* strains.

The fate of horizontally transferred sequences depends on their cost or benefit to the bacteria, with those that provide an obvious advantage under a wide variety of conditions expected to increase in frequency in the population and over long periods of time to evolve a base composition typical of the species, a process referred to as amelioration. The fact that we can recognise so many atypical sequences in the islands differentiating these closely related *E. coli* strains suggests that many of these horizontally transferred regions are temporary residents of the genome or provide an advantage specific to conditions encountered by particular strains.

E. coligenomes

E. coli K-12 and O157:H7 are more closely related to each other than either is to this extraintestinal strain. Again, although the island contents differ, we find many of the same chromosomal sites reused to house these strain-specific elements. Among the most notable observations is the lack of a type III secretion system encoded in this genome, given the apparent ubiquity of these systems in Gram-negative pathogens. There are, however, a number of previously unknown toxins and adhesions to fuel further research on the molecular basis of pathogenesis in the human urogenital tract. There is an ongoing project to sequence the genome of RS218, a quite different pathogen closely related to CFT073.

RS218 is associated with neonatal septis and meningitis, and comparisons with existing sequences should reveal genes common to both extraintestinal pathogens, but not found in diarrhoeagenic or commensal strains, and identify genes responsible for distinct aspects of their diseases. Molecular phylogenetics has clearly demonstrated, and it is now widely accepted, that Shigella are members of the species *E. coli*. This is clearly borne out by comparisons of the S. flexneri sequences with the other sequenced genomes. However, one thing was quite different—the rate of genome rearrangement is much higher than observed in previous comparisons. There are at least 15 rearrangements between the two Shigella genomes, and nearly all inversions are bounded by insertion sequences which are abundant in these genomes. In contrast, among all previous comparisons, there is only one apparent inversion event bounded by cryptic prophage in one of the two O157:H7 genomes.

There is no reason to think that interest in sequencing additional *E. coli* genomes is waning. There are a number of important pathogen lineages, such as those with a K1 capsular polysaccharide, enteroaggregative *E. coli*, enteropathogenic *E. coli*, and other Shigella species with sequencing projects underway and other significant groups remaining. The existing data raise important questions about the levels and mechanisms generating and maintaining genome variability within and between *E. coli* populations that will require substantial genomescale sequence information to resolve.

The abundant genetic, biochemical and physiological data for *E. coli*, emerging results from high-throughput microarray and proteomic experiments and computational models of whole-cell physiology promise future development of this system. By leveraging the strength of *E. coli* as a model organism and comparing processes across strains of varying levels of relatedness and different lifestyles we can adopt *E. coli* as a 'model bacterial species' where we can begin to dissect the complex processes involved in diversification of bacterial species.

BACTERIA AND DNA REPAIR

DNA repair now sits centre stage, along with DNA replication and the cell cycle, as one of those key processes essential to the function of every living cell. It is of course inaccurate to describe it as a single process—it is a comprehensive suite of functions, able to deal with almost every environmental challenge thrown at our DNA, maintaining the molecule's structure and genetic integrity.

Defects in repair impact on survival, mutation, cancer, antibody production, reproduction, meiosis, mitosis ...the list is endless. And yet, not too long ago, DNA repair was considered an odd little corner of biology, largely studied in a rather oblique manner by the 'radiobiologists'.

The subject was generally outside mainstream interests, and indeed it could be considered that a general realisation of the importance of repair processes really only came with the demonstration of the unequivocal link between mismatch repair and colon cancer in the 1980s. Almost no biochemical analysis of repair functions was even attempted until the 1960s, although good genetics had begun to indicate the presence of these processes somewhat earlier.

Repair in Microbes

The Haynes and Hanawalt suggestion has to be set in the context of the elegant genetic and physiological work, done primarily with bacteria and bacteriophages, but also with eukaryotic micro-organisms, which showed beyond doubt that DNA repair was occurring. And more, much more than this—the bacteria did it their way! First, there was lightdependent repair, or photoreactivation, discovered by Albert Kelner in 'the fungus' Streptomyces griseus. Then excision repair, discovered initially through the phenomenon known as 'host-cell reactivation', in which the damage done to UV-irradiated bacteriophages could be repaired upon infection into repairproficient bacterial cells.

The demonstration that host-cell reactivation was truly due to a repair process required one key ingredient a mutant bacterial strain in which repair was defective. This key ingredient came from the laboratory of Ruth Hill, with her isolation of the UV-sensitive bacterial strain Escherichia coli Bs^{-1}. This strain, some 100 times more sensitive to UV light than the parental strain, provided the vital tool with which to investigate the putative process of excision repair, allowing as it did a direct comparison at the biochemical level of the differences between mutant and wild-type parent. Many labs had by this time predicted how excision repair might work, but it was Dick Setlow who provided the proof, demonstrating that damagecontaining oligonucleotides were released from highmolecular-mass DNA in wild-type *E. coli*, but not in the sensitive Bs-1 strain.

E. coli

Discoveries from *E. coli* now came thick and fast. Paul Howard-Flanders' group isolated the excision-repair defective uvr mutants in *E. coli* K12, allowing a thorough genetic analysis to be undertaken. Dean Rupp and Howard-Flanders showed that a second type of dark repair, termed post-replication repair, occurred in excision-repair-defective mutants. Mutants defective in recombination were isolated by John Clark, and shown to be defective in this newly identified repair process. Mismatch repair was discovered, this elegant mechanism exploiting delayed methylation of newly replicated DNA to direct repair of mismatched bases to the newly synthesised strand.

The concept of excision repair was extended to base excision repair (removal of a single damaged base rather than an oligonucleotide), a mechanism involving different initiating enzymes from those identified in the uvr mutants, but sharing many later steps. The uvr-based mechanism was now renamed 'nucleotide excision repair' to distinguish it from this new mechanism. Studies of mutation resulted in the isolation of a number of classes of 'non-mutatable mutants', confirming that mutation too was an enzymemediated process, and culminating in the proposal of error-prone trans-lesion synthesis put forward by Evelyn Witkin and Bryn Bridges. In each of these cases, we now have flesh added to the bones. Base excision repair includes a whole set of mechanisms for the removal of different types of damaged (or incorrect) bases from DNA.

Mismatch repair is also multifaceted, a family of processes recognising different types of mismatched bases and of such importance that closely related enzymes are conserved from bacteria to man. Equally stunning was the elucidation of the genetics and biochemistry of the 'SOS response', the environmentally responsive system controlling the expression of many DNA repair and recombination enzymes. A second controlled system, identified initially by John Cairns as the 'adaptive response', allowed the *E. coli* cell to respond to methylation damage. Characterisation of this system resulted in the identification of the alkyltransferases, proteins (not strictly enzymes) that could remove alkylation damage from DNA by transferring it to themselves and which, in the case of the Ada protein, also played a key role in the control of the response.

GENOMIC ANALYSIS METHODS

When all the contigs have been joined and all the sequence ambiguities have been ironed out, the time comes in any genome sequencing project when it is necessary to annotate the completed sequence. Even scientists who are not directly involved in large-scale sequencing may have an interest in genome annotation and analysis, because bacterial genome sequences are now often released in a preliminary form -finished but not annotated, or even unfinished—many months before they are published. Obviously, annotation is not a trivial undertaking, but rather one that requires the use of an extensive range of bioinformatic skills. The focus of the project moves from the 'wet lab' with its gels, sequencers and PCR machines to the 'dry lab' of hardware, software and algorithms.

The sequencing project could also be said to move from the hard facts (the exact DNA sequence) to the softer inferences: attempting to identify potential open reading frames (ORFs), to assign homologies from sequence similarities, to identify motifs and fingerprints, to study codon usage patterns, to find genes that have been acquired recently through lateral transfer, and so forth.

It may sound as though a lot of hand-waving is involved, but with the recent development of exquisitely sophisticated algorithms many of the heuristic elements have been removed. At first glance, a completed microbial genome sequence appears to be an unruly and undisciplined assortment of the four nucleotides. However, a careful analysis, combining a sound knowledge of microbial biochemistry with good computational assistance can provide a surprising insight into the macromolecular architecture of a completed genome.

GENOMICS, PHYLOGENETICS AND EPIDEMIOLOGY

If the history of genetics has taught us anything it is that each technological advance in data generation, from starch gel electrophoresis to PCR and automated DNA sequencing, has brought with it vast new amounts of information, and often new debates, about how genomes are organised and evolve, and how species are related. It is not difficult to predict that the first data revolution of the 21st century -the ability to sequence and compare the entire genomes of organisms—will likewise bring with it many new insights into the genetical history of life on earth.

Epidemiology is one science which will benefit from this flood of gene sequence data. Access to the complete genomes of microbial species will doubtless assist in the identification of the molecular factors which underpin fundamental biological properties like virulence and tropism, while on the host side genome data will allow us to determine the genetic basis of disease susceptibility and outcome. However, even though the availability of sequence data will no longer limit the type of questions asked, there are still challenges for those wishing to use it to understand the nature of infectious disease.

First, it will be necessary to implement rational sampling protocols, controlling for factors such as age, sex, geographical location and disease classification, so that the sequence data we have are as informative as possible. Second, it will be necessary to develop a new suite of analytical tools which can effectively process such an abundance of information. It is likely that many of these tools will have their roots firmly planted in evolutionary biology—a field that provides a natural theoretical framework by which to understand why genomes are as they are.

Phylogenetic Tree

Perhaps the most useful tool in the evolutionist's kit is the phylogenetic tree. The growing sophistication of methods of phylogenetic analysis along with their ever wider application is one of the most significant developments in evolutionary biology. Since their first appearance in the early 1960s, gradual improvements have been made in the mathematical models used to describe sequence evolution, the proper starting point for any phylogenetic analysis. Not only can the models available today allow bases to change at rates specific to the data in hand, but methods are now so advanced as to allow changes in base composition along lineages and for different sites along a sequence to evolve at different rates, both of which can have profound effects on the trees constructed.

Progress is also being made in allowing bases to evolve in a concerted fashion, as might be caused by protein or DNA/RNA secondary structure. Put together, these advances mean that the assumptions we make when we construct a phylogenetic tree are increasingly likely to mirror the evolutionary processes that produced the sequences in the first place. Furthermore, techniques of tree reconstruction are being developed which use gene position or gene order.

Once constructed, phylogenetic trees can be put to a variety of uses in epidemiology, including determining the source of particular epidemics, tracking their subsequent progress through populations and assessing whether certain strains are more often associated with outbreaks than others. More recently, phylogenetic methods, such as that of Nielsen & Yang, have been developed which allow the identification of individual codons subject to positive selection. Such techniques obviously have enormous potential in the identification of those sites of utmost functional importance.

Equally impressive is that phylogenetic trees can even provide an indication of rates of pathogen spread. The use of molecular phylogenies in this context stems from the realisation that different demographic histories (i.e. rates of population growth or decline) alter the rates at which lineages appear on the tree. As a simple rule, trees reconstructed from a (small) sample of lineages from a population which has maintained a constant size through time will tend to have more branching (coalescent) events towards the tips than at the root.

As the rate of population growth increases, branching events move from the tips towards the root, from which dynamic it is possible to estimate approximate growth rates. It is therefore possible to measure the growth characteristics of epidemics simply by studying the branching structure of phylogenetic trees. As a case in point, application of this method to HIV-1 sequence data revealed different rates of population growth for viral subtypes A and B, suggesting that there was an initial rapid spread of subtype B viruses through high risk populations in Europe and North America. It is clear that as theory develops further, so we will be able to infer more complex epidemiological histories simply by examining gene sequences.

Although we can construct phylogenetic trees with more accuracy than ever before, and can ask more far-reaching epidemiological questions, this in itself does not necessarily mean that we are putting together a more accurate picture of evolutionary relationships. The limitation of most phylogenetic analyses is that a tree is merely a hypothesis of how sequences and species might be related to each other—albeit generally a very good one—and that there are other, more complex, ways in which organisms can be connected.

For example, if a sequence has undergone recombination in its evolutionary history, a process which is common within and among many microbial species, then it may be more realistic to depict their evolutionary relationships as an intricate 'network' diagram in which sequences have multiple ancestors, rather than as the normal bifurcating tree with a single ancestral sequence at each node. It is equally clear that recombination in the form of lateral gene transfer has been widespread across very great evolutionary distances, a process which will greatly complicate the desire to reconstruct the tree of all cellular life and to classify organisms in a simple hierarchical manner.

Deciding whether evolutionary patterns are better represented as trees or networks is not only a taxonomic nicety, but it also has profound implications for understanding evolutionary mechanisms and predicting future trends. For example, loci exhibiting high rates of recombination, and which form network structures, will have a greater capacity to generate genetic variation and hence have more adaptive potential than those in which mutation is the only fuel for diversity, and so evolve in a tree-like manner. Such accelerated evolutionary change is of special importance in antigen-encoding genes or those that determine drug or vaccine resistance. A key phylogenetic question for the future is, therefore, which genes evolve in a simple tree-like manner and which show more complex structures, as this provides a simple window on evolutionary mechanism.

Despite the evident importance of networked evolution, most methods of phylogenetic analysis will draw a tree regardless of the underlying evolutionary process and in doing so will misrepresent evolutionary history. As an example, consider the relationships between various species of Neisseria bacteria reconstructed using the 16S rRNA gene, a gene often used in the phylogenetic analysis of microbial species. Although a simple neighbour-joining tree of ten of these species appears to be a reasonable interpretation of their phylogenetic relationships, it ignores the fact that many interspecific recombination events have occurred.

In these circumstances a network structure is a much more realistic reconstruction of their evolutionary history. It is perhaps ironic that as gene sequence data accumulates and phylogenetic trees become ever more common and powerful analytical tools, there is a growing realisation that these trees are not always the best representation of evolutionary history.

The ability to sequence and compare the complete genomes of organisms promises much to many areas of biological science. For evolutionary biologists these data will allow us to unravel the crossed wires of evolutionary relationships, while for epidemiologists they provide a major opportunity to learn why pathogens spread and cause disease. The challenge for bioinformatics, now a key component in both disciplines, is not only to devise more efficient computer programmes to analyse the abundance of gene sequence data, but also to infer more complex patterns of evolutionary relationships, and hence mechanisms, from these sequences.

HORIZONTAL TRANSFER OF GENES

Horizontal transfer of genes in bacteria has been studied for several years, especially with regard to extrachromosomal elements. Genomic sequencing is now providing increasing evidence for widespread exchange of chromosomal genes.

Mechanisms of Genetic Exchange

Conjugation

Conjugation has been well studied, especially in Escherichia coli and Pseudomonas aeruginosa. In these species, sex factor (F and FP) plasmids can become integrated into the chromosome by homologous recombination between an insertion sequence (IS) on the plasmid and a homologous IS on the chromosome. Subsequent pilus synthesis and initiation of singlestrand displacement replication then causes the transfer of half of the F plasmid followed by a large segment of chromosomal DNA, via the pilus, into a recipient cell. There, the second strand is synthesised and the acquired DNA is incorporated by homologous recombination. Amongst Gram-positive organisms, streptococci are the best studied and contain both conjugative plasmids and conjugative transposons.

Transformation

This mechanism, while not ubiquitous, has been well studied in both Gram-positive (Streptococcus pneumoniae and Bacillus subtilis) and Gramnegative (Haemophilus influenzae and Neisseria gonorrhoeae) bacteria. Several other species are known to have natural transformation systems, among them Acinetobacter (where transformation may be the major mechanism of dissemination of antibiotic resistance in this species) and the cyanobacterium Synechocystis. The completion of the genomic sequences of B. subtilis and H. influenzae and the identification of their competence genes makes it possible to search for homologues in other species.

Transduction

Two types of transduction exist: specialised (by aberrant excision of a lysogenic phage, incorporating genes adjacent to the phage attachment site) and generalised (by erroneous encapsidation of chromosomal DNA in lieu of linear phage concatameric DNA produced by rolling circle replication). The specificity is restrained by the phage host range; this is probably not a major mechanism of interspecies transfer.

Vehicles for Genetic Exchange

Plasmids. Plasmids have been best studied in *E. coli* and P. aeruginosa. Early on they were divided into F (sex factor) and R (resistance) plasmids. Later, many R plasmids were found to encode F-like replication and/or transfer genes. Conjugative R plasmids are responsible for most of the dissemination of antibiotic resistance genes. One of the best studied is R100, which contains F-like transfer genes but different replication genes. Within R100 there is a transposon, Tn10, encoding tetracycline resistance, and another transposon, Tn21, encoding mercury resistance, itself within a Tn9-like transposon encoding chloramphenicol resistance. Within Tn21 there is an integron encoding streptomycin and sulphonamide resistance.

While the aforementioned are typically extrachromosomal genes, plasmids can also carry chromosomal genes. A good example is the class C chromosomal b-lactamases, whose genes are increasingly found on plasmids. As mentioned above, F plasmids can transfer large blocks of chromosomal genes.

Transposons

Transposons, or 'jumping genes' are responsible for the dissemination of several antibiotic and heavy metal resistance genes, as well as degradative genes and even the lactose operon (the latter with the consequence that clinical microbiologists can no longer ignore Lac+ enterobacteria). Perhaps the best-known example is Tn3, which contains two genes for its own transposition and the TEM-1 blactamase gene. Tn3 has spread among Enterobacteriaceae on plasmids of several incompatibility groups (which correspond to replication functions and thus to host ranges).

But perhaps the best example of horizontal transfer by transposons is the arrival of Tn3 in Haemophilus and Neisseria in the mid-1970s. Although not all the pieces of the puzzle can be retraced, the most likely scenario would involve the transfer, by conjugation or transformation, of an *E. coli* plasmid carrying Tn3 into Haemophilus ducreyi (the agent of soft chancre). The plasmid would not have been able to replicate (*E. coli* plasmids are unknown in Haemophilus) but before it was degraded, its Tn3 'hopped' onto a native plasmid.

The H. ducreyi plasmids are the 'smoking gun' since they contain all of Tn3. A 'streamlined' version, having lost the transposase gene (no longer necessary since the plasmids are spread among Haemophilus and Neisseria by transformation), is found not only in H. ducreyi but also in Haemophilus parainfluenzae and N. gonorrhoeae, where it caused the emergence of the well known penicillinase-producing N. gonorrhoeae (PPNG). Surprisingly, the TEM gene in these organisms has not produced the wide variety of extended-spectrum TEM b-lactamases found in Enterobacteriaceae despite the widespread use of ceftriaxone to treat PPNG—does N. gonorrhoeae have a very faithful DNA polymerase?

Complex transposons, such as Tn5, are composed of a central section, typically containing one or more antibiotic resistance genes, flanked by inverted or direct repeats of ISs. These transposons are probably formed by transposition of ISs to either side of a region of DNA. Indeed, any region of chromosomal DNA which becomes flanked by copies of the same IS can potentially become a transposon. This should be looked for in sequence analysis of chromosomal DNA near putative IS elements.

Conjugative transposons are found in Gram-positive bacteria and encode the mechanism, yet to be described in detail, for their own transfer. They encode an integrase and an excisase related to those of temperate phages. Among these are Tn916, encoding tetracycline resistance, Tn1545, encoding erythromycin resistance, and Tn1547, encoding one type of vancomycin resistance (another is carried on Tn1546, a Tn3-type transposon).

Integrons

Much of the dissemination of antibiotic resistance in Gram-negative bacteria is due to integrons. Evidence for their existence was hinted at by restriction maps and heteroduplexes, demonstrated by electron microscopy, which pointed to gene-sized differences among related plasmids and transposons. Sequencing confirmed that a mechanism of site-specific recombination, mediated by an integrase, was at work.

Integrons are elements in which structural genes, each linked to a palindromic '59-base element' (attC site), are assembled as cassettes into operons by insertion at an attachment (attI) site. A strong promoter adjacent to the attI site makes integrons a sort of natural expression vector. In class 1 integrons, the most frequent kind, the integrase and attI site (from a cryptic phage in chromosomal DNA?) became associated with a Tn5053-like transposon, and a few have retained this association, e.g. Tn402 of plasmid R751.

Similarly, class 2 integrons became associated with an ancestral transposon to form Tn7, whose integron carries its resistance genes. Most elements carrying class 1 integrons are descended from an element which acquired sulfonamide resistance, but in turn lost two of the four transposition genes. However, one of these was able to regain mobility by inserting into a mercury resistance transposon, Tn2613, to form Tn21. Many different cassette arrangements are found in Tn21-like transposons.

Although no direct chromosomal ancestor of class 1 or 2 integrons has yet been found, genomic sequencing has revealed several new classes of integrons, indicating that they are ancient mechanisms of gene exchange which have only recently been co-opted for dissemination of antibiotic resistance. In Vibrio cholerae there is a 'super-integron' in which over 150 cassettes (mostly unidentified genes), each with a palindromic 'VCR repeat' (a variant attC site) are aligned in a row next to an integrase gene.

Other classes of integrons exist in Xanthomonas campestris pv. badrii, Shewanella putrefaciens, and in Nitrosomonas europaea. The attC sites in their cassettes are much more similar to the '59-base elements' of class 1 integrons than are VCR repeats. The integrases of the various classes, however, all show 45-60 % amino acid identity between themselves.

STRUCTURAL PATHOGENOMICS

Structural genomics uses the information gleaned from genome sequencing to determine structures of the gene products. A 'structural genome' implies a complete 3-D description of every protein in an organism. Experimentally, this will prove difficult, as each protein is unique—unlike DNA, which is tractable for sequence analysis regardless of its source. However, proteins can be described by a finite number of representative structural 'folds' and useful 3-D models can be built if the protein shares a fold of known structure.

This promises a structural metagenome, where all protein structures can be modelled. The huge task of generating the fold library requires a multidisciplinary effort, which has led to a number of consortia that collaborate to churn out structures by nuclear magnetic resonance, X-ray crystallographic and computer modelling techniques. The structural genomics community has dedicated a lot of effort to the development of high-throughput, automated techniques for the workflow, and some groups focus on the analysis of pathogenic organisms.

New Antimicrobials

As well as identifying proteins implicated in pathogenesis, the analysis of genome sequences can suggest specific metabolic pathways and enzymes representing weak points in the pathogen's defence and survival strategies. Proteins essential for the viability of the pathogen, which are absent or significantly different to their human counterparts, are likely to be good candidates for antimicrobials. As an example, the causative agents of malaria and tuberculosis (Mycobacterium tuberculosis), together responsible for an estimated 5 million deaths per year, share a fatty acid biosynthetic pathway that is not found in humans. Screening has identified compounds that are active against a key enzyme, enoyl reductase, in this metabolic pathway. The X-ray structure of the Mycobacterium enzyme with one of these compounds elucidates the binding characteristics that can be further exploited for drug optimisation and so provide a powerful tool in fighting both malaria and tuberculosis.

Structural genomics is also providing structures of proteins that identify new targets for drug intervention. These include essential proteins of currently unknown function, and proteins whose biological functions have been illuminated by the structure. For example, the Joint Centre for Structural Genomics in California has uncovered thymidylate synthase complementing protein (TSCP) as a potential drug target. Blocking TSCP prevents bacteria from synthesising new DNA, leaving the human thymidylate synthase enzyme unaffected.

Structure-based Drug Discovery

The high-resolution structure to 1.6 Å of the TSCP enzyme has allowed the arrangement of the active site to be investigated at a level of detail that would not be possible with a modelled structure. This precision makes knowledge-based drug design a powerful tool.

Small molecules likely to bind to the protein can be tested in an experimental or a virtual way. Computer-based in silico screening involves large chemical libraries and a variety of docking or scoring tools being applied to find favourable interacting partners with the protein target. The virtual approach has, of course, to be verified experimentally, by cocrystallisation of candidate compound and protein, and eventually by testing the inhibitory effect in living cells.

Prokaryotic and Eukaryotic Microbes

Prokaryotic organisms face a similar challenge to simple eukaryotes in the event of damage to their DNA. As each prokaryotic cell is destined to divide, the focus of the DNA repair and damage response is in protecting the integrity of the genome. In bacteria DNA damage and repair are continually monitored and a primitive damage response, termed the SOS response, is activated to co-ordinate repair and cell division. Emergency repair pathways of nucleotide excision repair, homologous recombination and lesion bypass are induced as part of the SOS response, in parallel with temporary cessation of cell division.

Once the DNA is repaired, the emergency repair systems are turned off and cell division is allowed to proceed. Remarkably, the mechanisms of DNA repair processes are essentially conserved from prokaryotes, through simple eukaryotes to complex multicellular organisms, such as man. In many cases, pivotal repair enzymes show a high level of conservation of sequence and structure from bacteria, through simple eukaryotes, such as yeast, to man. In view of the increased complexity of eukaryotic organisms it is perhaps not surprising, however, that the complexity of the enzymatic machinery that mediates repair is increased approximately 10-fold between bacteria and man.

GENETIC IDENTIFICATION OF MICROORGANISMS

The genetic identification of microorganisms utilises molecular technologies to evaluate specific regions of the genome and uniquely determine to which genus, species, or strain a microorganism belongs.

Genetic identification of microorganisms is basically a comparison study. To identify an unknown organism, appropriate sequences from the unknown are compared to documented sequences from known organisms. Homology between the sequences results in a positive test. An exact match will occur when the two organisms are the same. Related individuals have genetic material that is identical for some regions and dissimilar for others. Unrelated individuals will have significant differences in the sequences being evaluated.

Developing a database of key sequences that are unique to and characteristic of a series of known organisms facilitates this type of analysis. The sequences utilised fall into two different categories, 1) fragments derived from the transcriptionally active, coding regions of the genome, and, 2) fragments present in inactive, noncoding regions. Of the two, the noncoding genomic material is more susceptible to mutation and will therefore show a higher degree of variability.

Depending on the level of specificity required, an assay can provide information on the genus, species, and/or strain of a microorganism. The most basic type of identification is classification to a genus. Although this general identification does not discriminate between the related species that comprise the genus, it can be useful in a variety of situations.

ORGANELLES AND SUBCELLULAR GENETICS

Organelles are internal cellular structures that perform dedicated functions. Oraganelles includes tructures such as ribosomes, mitochondria, chloroplasts (the site of photosynthesis in plants and other photosynthesising organisms), endoplasmic reticulum (ER), Golgi apparatus, and lysosomes.

The mitochondrion of all eukaryotes and the chloroplasts of plant cells are the only organelles that have their distinct genomes. These genomes are made of a single, circular DNA molecule denoted mtDNA in mitochondrion and ctDNA in chloroplast. The replication and the mode of inheritance of organelle genomes are distinct from the nuclear genomes.

Mitochondrial genomes vary in size, among species, by up to one order of magnitude. Animal cells have a genome of approximately 16 kb (kilobases, 1,000 bases) and represent the smallest mitochondrial genomes in eukaryotes. Yeast possess a much larger genome that varies among the different strains but is about 80 kb, with the whole yeast mitochondrial DNA making up 18% of the total DNA of the yeast. Plant mitochondrial genomes are the largest and most complex. They show an extremely wide range of variation in DNA size. The smallest plant mitochondrial genome is around 100 kb, which make it very difficult to isolate in intact form. These genomes contain short homologous sequences that may undergo recombination thus generating small circular molecules that coexist with the intact ctDNA.

Eukaryotic cells may contain up to several hundred mitochondria. These mitochondria contain their own replication, transcription and translation systems. Nuclear genes, however, encode the majority of mitochondrial proteins which are synthesised in the cytosol and then targeted to the mitochondrion. Each mitochondrion can contain up to ten copies of the circular genome. A special DNA polymerase replicates the mitochondrial genome. The complete sequencing and mapping of several mammalian mitochondrial genomes show extensive similarity in organisation.

The mammalian mitochondrial genome is extremely compact with many overlapping genes and no introns. This genome codes for 13 essential genes of biochemical pathways (e.g., oxidative phosphorylation), two rRNAs, and 22 tRNAs. The genetic code of mitochondria differs from the standard genetic code used by the cytosolic ribosomes and the bacterial ribosomes. There at least two codons for all the aminoacids and plus four termination codons. The yeast mitochondrial genome is much larger, but codes for only eight proteins. The mitochondrial products synthesised by this genome, both RNAs and proteins, are similar to those produced by the mammalian mitochondria. The most distinguishing feature of the yeast genome is the existence of interrupted loci. The introns in some are so large that their size is almost as large as the whole mammalian mitochondrial DNA.

The genomes of the chloroplast of different plant cells are relatively large but show considerable difference in overall length, between 100 kb and 200 kb. The complete sequence and mapping has been determined for some organisms. These sequences show a highly conserved overall gene number and organisation in the different species. These genomes usually encode for 50–100 proteins as well as rRNAs and t RNAs. The majority of the characterised proteins are involved in gene expression, electron transfer or in photosynthesis. The latter form complexes located in the thylakoid membranes. Both protein-coding genes and those coding for tRNAs contain introns.

Both mitochondrial and chloroplast genomes are believed to have evolved through endosymbiosis. This model of organelle evolution proposes that eukaryotic cells captured bacteria that later provided the function of mitochondria and chloroplast. Phylogenetic studies based on DNA sequence analysis suggest that mitochondria and chloroplasts evolved separately from eubacterial lineages related to purple bacteria and cyanobacteria, respectively. Mitochondrion is presumed to have evolved from a species that is very similar to *Rickettsia*, an obligate the intracellular bacteria.

Both mitochondria and chloroplasts follow a non-Mendelian mode of inheritance usually referred to as extranuclear or cytoplasmic inheritance. Most mtDNA is inherited from egg cells and thus show a maternal pattern of inheritance. This has implications both in studies of evolution between different populations of species as well as in the inheritance of some genetic diseases.

Because of the relative simplicity of the mitochondrial genome, genetic manipulation has advanced considerably. The first humans with half the nuclear genome from one mother and the mitochondrial genome from a donor mother have been delivered and are reported healthy. This genetic manipulation has now been carried out on about thirty children in the United States. Technically, this amounts to a kind of germline genetic modification and has, therefore, caused great reservation among the international biomedical community. Many scientists and physicians argue that there are serious ethical issues that should be addressed before such a practice should be allowed.

REFERENCES

Casadesús, J. & R. D'Ari, "Memory in bacteria and phage", *BioEssays*, 2002.

GreA and DksA. Potrykus, K., D. Vinella, H. Murphy, A. Szalewska-Palasz, R. D'Ari & M. Cashel, "Antagonistic regulation of *E. coli* ribosomal RNA *rrnB* P1 promoter activity" Journal of Biological Chemistry, 2006.

Vinella, D., C. Albrecht, M. Cashel & D'Ari, R., "Iron limitation induces SpoT-dependent accumulation of ppGpp in *Escherichia coli*", Molecular Microbiology, 2005.

4

Medical Microbiology

The evolution of medical microbiology is a response to clinical needs. This can take the form of a technical innovation garnered from the armamentarium of science in general or a reflection of advances in anti-infective therapy demanding recognition of microbial etiologies that now respond to therapeutic intervention. But this history is also a reflection of the political climate, of the perceived threats posed by an emerging field to more established organizational and professional entities. While this struggle greatly influenced the development of clinical microbiology as a distinct specialty, it is sufficiently controversial to be omitted from this essay.

Clinical microbiology got its start when various stains became available to indicate the presence of different organisms. The Gram stain helped divide the vast array of bacteria in various specimens into categories based, in addition to the staining reaction, on the anatomy of the organism and its source. The acid-fast stain aided in the recognition of mycobacteria, while the Albert stain suggested the presence of Corynebacterium diphtheriae, leading to the administration of a specific antiserum, an emerging therapy at the start of the 20th century that was usually used on the basis of clinical presentation.

A variety of selective, supplemented, and enriched culture media became available for isolation, permitting more rapid recognition of the presumptive presence of a significant organism. Antisera were put to use as therapies and to identify isolates. The Quellung reaction became a standard task of interns and resident house officers, performed to identify the type of pneumococcus present and guide specific serum therapy, in addition to avoiding serum sickness by obtaining careful histories of previous treatment with equine antibodies against other etiologic agents.

The advances in the grouping and typing of streptococci, salmonellae, and shigellae, the separation of Staphylococcus aureus on the basis of the coagulase reaction, and the growing awareness of the need for safe water and uncontaminated food items established the need for laboratories to assume these responsibilities. It was only logical that microbiology should join endeavors such as chemistry, hematology, and serology under the rubric of clinical pathology.

Differential media especially designed to sequester species increased dramatically during Word War II; military hospitals developed clinical microbiology sections devoted not only to recognizing agents endangering the health of troops in camps, in battle, and in foreign environments but also to assessing the responses of certain of the microorganisms isolated to several sulfonamides and that hitherto unknown agent, penicillin. The subsequent explosion of antimicrobial agents-streptomycin, chloramphenicol, tetracyclines, and erythromycin-suggested to the reigning powers of medical facilities that clinical microbiologists could be phased out, since infectious disease would disappear before the onslaught of agents discovered through human ingenuity.

In the interim, cotton plugs gave way to Bakelite, polypropylene, glass, metal, and plastic closures; in-house medium preparation was relieved in part by the beginnings of commercially manufactured ready-to-use media especially for mycobacteria and antimicrobial susceptibility testing. Alcohol, Bunsen, and Tyril burners were replaced by microincinerators, eventually followed by disposable loops and transfer needles.

The prescient wisdom of hospital boards soon was shattered by the genetic versatility of the microbial world, dramatically demonstrated by the pandemic of S. aureus 80/81 in the late 1950s and early 1960s and the emergence of gram-negative rods that demonstrated the superiority of the bacterial physiology over the commercially prepared secondary microbial metabolites that initially appeared so promising. To be sure, the tug of war between antimicrobial agents-natural and synthetic -and the microorganisms continues unabated, with signs that the evolutionary potential of the microbial world will succeed in the long run.

Since the 1960s, numerous ingenious innovations have been introduced and used in clinical microbiology laboratories. Parenthetically, none of them allowed microbiologists to abandon the dogma of the pure culture technique, to escape the need to study microorganisms under the most unnatural conditions, or to work with the laboratory variants of the organisms encountered in the natural setting. Molecular biology techniques promise to revolutionize the diagnosis of infectious disease -to date a promise still in its infancy.

Systems approaches began to replace the single test tube with but one substrate. Perhaps the first was double sugar iron agar for the recognition of so-called enteric pathogens, followed by triple sugar iron agar and the next tentative shortcut, the r/b tube. Rollender and Beckford, the inventors of the r/b tube, must be credited with initiating manufacturers' efforts to teach laboratory staffs the vagaries and problems of new system approaches. Shortly thereafter, the API system was introduced in the United States, bringing a novel numerical approach first to the identification of Enterobacteriaceae and then to that of several other categories of microorganisms.

Similarly, the Roche Enterotube used fewer reaction substrates to decrease the time needed to identify isolates to the species level; initially it was used for members of the Enterobacteriaceae and eventually for other microbial representatives. All systems eventually addressed yeasts and nutritionally demanding bacteria, obviating the multiple-tube approaches in use. Miniaturization, plastic disposables, and commercial medium production are the norm in present-day microbiology laboratories.

Mechanization of analytical microbiological procedures gave way to automation. The first automated procedure was Technicon Automated Antibiotic Susceptibility (TAAS), followed by variations of the TAAS approach such as Autobac (Pfizer) and MS2 (Abbott). Blood cultures were and are still addressed by BacTek, also used for mycobacterial susceptibilities. The McDonnel-Douglas system for detection of life on Mars was modified to the Automated Microbiology System (AMS) for use in the clinical microbiology setting, now a product of bioMe´rieux after a Vitek interval; it has been followed by a growing number of various machines performing the tasks of yesterday's technical staffs. Robotization, now available for the routines of chemistry, hematology, endocrinology, some serology, and urinalysis, may eventually find applications in some if not many aspects of clinical microbiology.

But the very nature of clinical microbiology has undergone profound changes in the past few decades. Hitherto unknown microorganisms and viruses have appeared, often as the result of human activities with environmental impacts. Etiological agents unknown locally may be introduced by the faster transportation of people, manufactured goods, and foods, wreaking havoc among immunologically innocent populations. The task of identifying, isolating, and evaluating therapies for these emerging host-harmful entities has become part and parcel of the present function of clinical microbiologists, aided frequently by the tools of molecular biology. The threat of bioterrorism adds a new dimension to the steadily growing number of emerging infectious agents.

The modalities and procedures of clinical clinical microbiology are intimately entwined with advances in the diagnosis and treatment of infectious diseases. The pressure for rapid and accurate diagnosis is immeasurably enhanced when therapeutic measures for a specific disease come into use. Thus, the impetus to improve the microbiological acumen was felt as early as the era of serotherapy, when that acumen was needed to monitor vaccine effectiveness and to identify agents of sexually transmitted diseases for early chemotherapeutic measures, followed by the era of antimicrobial agents and, more recently-especially spurred on by the human immunodeficiency virus-the era of antiviral and antiparasitic compounds.

Clinical microbiologists are acutely aware of the constantly emerging intruders into the intimate human biosphere. These agents appear as the traditional scourges of humanity are brought under control. But the application of antimicrobial agents to the food chain, cosmetics, and over-the-counter medications, and the advances in medical science, sparing individuals afflicted with a variety of diseases but accompanied by impaired immunity-all these factors have combined to increase nosocomial infections, placing the medical facility at the very apex of the selective-pressure pyramid.

The selection results in colonization by microbiota with a minority of antimicrobial-tolerant or-resistant constituents; administration of antimicrobial therapy converts these organisms to a majority. These selected prokaryotes and eukaryotes, along with the emerging viruses, coccidia, yeasts, and molds, pose a dynamic challenge to the clinical microbiologist and promise a continued need for her or his services. But these challenges must be met by the expansion of technical skills brought to bear on the changing nature of the challenging microbiota and the willingness of clinical microbiologists to adopt and practice evolving technologies, to gain knowledge in addition to information, and to remain in the forefront of innovation and invention.

DEVELOPMENT OF ANIMAL VACCINE

Veterinary vaccines can be used to protect animal health, but by preventing zoonotic infections animal vaccination also protects human health, as exemplified by wildlife vaccination against rabies. In animal health the focus is now on animal infections rather than on animal diseases. Vaccines should be designed to prevent infection rather than to prevent clinical signs of disease and should, wherever possible, produce sterile immunity. Last but not least, available technologies allow us to design DIVA vaccines, together with their companion diagnostic tests, which make it possible to distinguish between vaccinated and infected animals even if the latter were previously vaccinated.

Smallpox, a human disease, was the first viral infection to be eradicated; eradication means the complete elimination of the disease and its infectious agent worldwide. This remarkable success was due to several factors, including the availability of an efficacious vaccine, namely vaccinia, and the absence of a wildlife reservoir. According to the World Health Organization (WHO), eradication of human poliomyelitis and measles may also soon be achievable for the same reasons.

The only animal virus disease for which the same circumstances exist is rinderpest: there are several efficacious vaccines already available and the infection seems to reach a dead-end if transmitted to susceptible wild species. The same cannot be said for other animal viral infections, none of which are likely to be eliminated in the near future, either due to the lack of an efficacious vaccine (e.g. African swine fever) or to the existence of wildlife reservoirs such as the wild boar (*Sus scrofa*) for classical swine fever or the African buffalo (*Syncerus caffer*) for foot and mouth disease. These diseases are more prone to regional elimination than to complete eradication worldwide. The choice of method to eliminate an animal infectious disease must take into account the biological and epidemiological characteristics of the infection, the available control techniques and the emergence of new vaccination technologies such as DIVA vaccines.

The reasons to develop veterinary vaccines are now manifold:

— to protect animal health
— to eliminate/eradicate an infection
— to improve animal welfare
— to protect public health
— to protect consumers from certain risks which may be linked to products derived from food-producing animals
— to protect the environment and biodiversity
— to avoid the emergence of pathogens resistant to available drugs
— to promote sustainable agriculture and food-producing animal production.

Unfortunately, even though the reasons for developing veterinary vaccines are many, there are still many obstacles to their development:

— scientific obstacles, such as those that prevent the development of vaccine for African swine fever, theileriosis, and many parasitic diseases

— difficulty in accessing the target species (wildlife)
— poor investment return for companies involved in vaccine development and production
— animal health regulations that prohibit the use of vaccination
— regulatory requirements for vaccine registration
— the so-called 'minor' species status of some targets
— conditions of minor importance in so-called 'major' species
— conditions of minor importance in so-called 'minor' species (the worst-case scenario).

The use of veterinary vaccines has obvious benefits for animal welfare. Vaccines, unlike therapeutic treatments, are the best way of avoiding animal suffering since they prevent disease or avoid the need for slaughtering as part of the implementation of stamping out. Furthermore, due to the short lifespan of many food-producing animals, vaccine need only be administered once, while treatments generally necessitate repeated interventions. Nevertheless, there is still room for improvement by developing less reactogenic adjuvanted vaccines. Another area of animal welfare improvement is the use of vaccines for immunocastration of male pigs to avoid boar taint, instead of surgical castration.

The use of vaccines in animal production systems is also often more environmentally friendly since it reduces the use of chemicals. Of special interest is the anti-tick vaccine developed in Australia, which is based on a cryptic intestinal antigen of the parasite. One should also mention the trials carried out in Australia to reduce methane (a greenhouse gas) emission by ruminants by vaccinating them against *Archeobacteria* of the rumen, although unfortunately this has had little success as yet.

VETERINARY VACCINATION AND HUMAN VACCINATION

Veterinary vaccination differs a priori from human vaccination in terms of the ethical issues surrounding experimentation, and the importance, and even the priority, of economic considerations when it comes to animal health. There is also a major difference in the use of alternative solutions to the vaccination or treatment of sick livestock, such as mass culling, a strategy often employed in veterinary public health, despite the high cost and the shocking image it creates. No farmer can remain indifferent to having to have his livestock culled, especially if they are healthy. And even the wholesale destruction of mere battery chickens is not an operation to be treated lightly.

Anyone who has tried to save a bird caught in an oil slick will find it hard to accept that in the 21st Century the only way of avoiding an epizootic is to destroy entire populations of poultry, cows or sheep, or even stray dogs. Yet however unique it may be in many of its theoretical or practical aspects, veterinary or animal vaccination has a scientific history that is closely linked to that of human vaccination, for which it has served as a model, a tutor and a complement. This proximity and even interconnection illustrate how in many ways veterinary medicine offers a wealth of observations unmatched by human medicine, confined as it is to the anatomy and physiology of Homo sapiens.

Not to mention that there are many human diseases where the reservoir is found in animals (rabies, for example), that the species barrier to infections is often crossed, and that many epizootic diseases prove to be potentially dangerous for humans, as in the case of avian influenza, all of which indicates the need for close collaboration in research.

Both human and veterinary medicine have certainly found a source of inspiration in the long tradition of empirical procedures where farmers have used fluids from sick animals to protect their herds. Several attempts at immunisation by inoculation were made for sheep pox, which is close to smallpox in humans, and bovine contagious pleuropneumonia. For the latter disease, the Belgian physician Willems brought this age-old practice into the scientific era when, from 1853, he inoculated animals at the base of the tail with a small amount of infective material.

The tissues, and no longer just the 'humours', then came to be studied under the microscope and underwent all kinds of procedures to try to achieve a permanent, stable attenuation or neutralisation. The telling observations of animal farmers and veterinary practices thus provided the historical crucible for contemporary vaccinology. The best example of the close relationship between human and animal vaccination, and certainly the best documented, is the history of successive vaccines against smallpox.

Vaccination against smallpox: an example of the historic links between human vaccine and animal vaccine. The inoculation of serous fluid under the skin is a procedure that has long been known as a way of protecting flocks against sheep pox. In particular, there is documentary evidence of its use by nomadic herders in Africa, for example among the Tulani. There can be no doubt that this practice must have drawn attention to the possibility of acquiring protection from a serious disease by contracting a form of the disease that was attenuated to a greater or lesser extent. In earlier times people were closer to their animals, and animal farmers often had the reputation among neighbouring townsfolk of being healers. Yet it is difficult to know whether it was inoculation with sheep pox that led to the idea of human variolation or vice versa.

The history of the vaccine against smallpox, a human disease with no known animal reservoir, can be summed up as the replacement of inoculation with human smallpox with inoculation with cowpox, a procedure invented by an

English doctor, Edward Jenner (1749-1823). The use of cowpox is generally seen as a remarkable advance compared to variolation. The latter technique used only human material, serous matter from pustules and scabs taken from a subject with a mild form of the disease. It generally conferred solid immunity. However, the outcome was unpredictable and postinoculation mortality was not inconsiderable.

In contrast, inoculation with cowpox, proposed by Jenner in 1798, seemed to be less dangerous and just as effective. Through a form of cross-immunity it provided humans with satisfactory protection, though probably less solid than that produced by the inoculation of smallpox. Indeed, during the 19th Century it proved necessary to revaccinate in order to reactivate the immunity since this tended to decline over the years. The need to revaccinate complicated the task of the health services and met with incomprehension on the part of the public, obliged to repeat a procedure that they had been led to believe was permanent.

At the beginning of the 19th Century, Jenner's vaccination procedure rapidly spread around the world, supported by governments favourable to a measure that could reduce the devastating effects of epidemics on their populations. The President of the United States of America (USA); the Tsar of Russia; the King of Sweden; the Emperor of France, Napoleon I; and the Pasha of Egypt, Ali Mohammed, to mention but a few, were greatly enthusiastic about the vaccine and actively promulgated it, in some cases, as with Napoleon I in 1812, going as far as to make it compulsory in the army, and even in society as a whole. When it came to putting these plans into action, however, it was of course quite a different story. Yet it was not long before vaccination with animal vaccine underwent changes.

In fact the use of lymph of animal origin that was subsequently 'humanised' soon became established: the original vaccine, derived from a cow, was first propagated from arm to arm, usually in children, who were used as vaccinifers. The method raised numerous problems. The lymph eventually lost its potency and produced hardly any pustules. Parents were also reluctant to have their offspring used as a reservoir for producing vaccine. Lastly, because repeated samples were taken from the same pustules they were soon emptied of smallpox virus, either because the pustules dried up or because they became superinfected and they then produced a fluid of dubious content.

During the latter half of the 19th Century, it seemed more natural and more practical to go back to the original source, namely cows or indeed calves, which were the only means of obtaining an authentic cowpox vaccine and ensuring an abundant and readily available supply of lymph. This had to be organised in a completely different way. Breeding centres had to be established for animal vaccinifers, and animals had to be transported from village to village by road or rail or else the vaccinal lymph itself had to be transported, kept in an appropriate medium to conserve it and protect it from superinfection. Various excipients such as liquid paraffin, lanolin and glycerine were tested, with glycerine eventually being preferred. As well as causing specific organisational problems, the transition from a vaccine of human origin to a vaccine of animal origin met with socio-cultural problems.

In India, for example, where smallpox was a terrible scourge for the densely populated continent, variolation was an ancient tradition dating back to at least the 17th Century. Throughout the 19th Century, the British colonial administration went to great lengths to develop the vaccine but had to contend with the reluctance of Hindus. They found the use of sacred animals for this purpose abhorrent, and the presence of a fatty excipient led them to suspect the use of animal fat prohibited by culture. Furthermore, the highest castes commonly practised smallpox inoculation which, as in England in the 18th Century, was accompanied by a set of dietary measures and isolation of inoculated subjects that were considered quite satisfactory.

Compared to the results of these inoculations, the variations in efficacy of inoculation with vaccine lymph were sometimes far from convincing. For a long time, the British administration steered a delicate course by using vaccination only for mass campaigns among the lower social classes. Vaccination, as a mark of solidarity against contagion, was thus confronted by the imperviousness of the caste barriers within Indian society. Yet, in the case of smallpox vaccine, can one legitimately call it an animal vaccine? Right from the start, when the Jennerian procedure was first disseminated, it became difficult to determine the exact origin of the vaccine being used. In England, the practice of vaccination in hospitals formerly used for smallpox inoculation took place without the subjects being isolated, and was accompanied by a hybridisation of strains.

Moreover, in countries that were in favour of vaccination it seemed preferable to identify and use local cases of cowpox rather than having to rely on a supply from abroad. Yet in many countries in Africa (e.g. Egypt in the 1830s) or Asia, spontaneous cowpox could not be found. An alternative solution that was tried in India was to inject cows with human smallpox in the hope of obtaining an unlimited supply of attenuated material. Small institutes, in Bombay, for instance, bred calves and produced stocks of lymph for distribution to villages. The lymph was injected into subjects and then transferred from arm to arm. Children were targeted first because of their lower susceptibility, and served as 'guinea pigs' to standardise the vaccine fluid (according to the number and appearance of the pustules produced), before its use in adults. Parallel controls to test for innocuousness were sometimes performed using donkeys or rabbits.

In the history of smallpox vaccination, it is therefore very difficult to distinguish between the paths of these two fluids, one human and one animal. We can do no more than speculate about the origins of the strains that we have today. It seems likely that Jenner's original strain has been irremediably lost. Three types of virus are commonly distinguished, according to the type of cell lesions in the culture media (embryonated egg or allantoic membrane): 'historic' cowpox virus (thought to be closely related to the strain used by Jenner), vaccinia virus, and 'classic' smallpox virus. However, it seems likely that what we have today are in fact intermediate strains.

Virologists are currently discussing a possible link between the smallpox vaccine and an equine virus that no longer exists in the wild. We have therefore eradicated smallpox before fully elucidating the origin and behaviour of poxviruses and their vaccines throughout history. The development of a new vaccine, free from the dangers of its predecessor, to protect against any future use in bioterrorism, will probably not help us to learn more about the past. The vaccine against smallpox, despite its many particularities, served as the inspiration for the development of vaccination against other diseases and as a springboard for the Pasteurian programme sometimes summed up as 'une maladie, un vaccin' ('for each disease, a vaccine').

VETERINARY VACCINOLOGY

Rinderpest

Rinderpest is one of the great historic plagues that have ravaged human livestock for centuries. In Europe, rinderpest was the major plague of cattle up to the end of the 19th Century, when it was eliminated. At about the same time rinderpest was introduced with devastating effects in Africa where it decimated the cattle and buffalo populations (Syncerus caffer), along with those of other susceptible domestic ruminants and many wildlife species. It is remarkable that rinderpest was eliminated from Europe by the end of the 19th Century by the simple application of sanitary measures, before the nature of the infectious agent was known.

In fact, the ability to control rinderpest effectively was often considered to be a measure of the quality of a country's veterinary services. When rinderpest was reintroduced in Belgium in 1920, it was again eliminated purely by sanitary measures within seven months and without spread to neighbouring countries. The history of medical prophylaxis (vaccination) against rinderpest illustrates the evolution of medical thinking. The Italian Lancisi wrote very lucidly: 'if such a dreadful disease were to threaten our cattle, I would be in favour of destroying all sick or suspect animals, rather than allowing the contagion to increase, simply to gain time in the hope of achieving the honour of discovering a specific remedy, which is often a fruitless quest...'. His contemporaries did not necessarily share this view and Ramazzini was convinced that rinderpest, being a disease similar to smallpox, could be controlled by homologous inoculation. This lead to a whole series of unsuccessful inoculation trials.

Some clearly stated that inoculation should only be recommended for areas already contaminated, otherwise there was a risk of spreading the disease further. Among all the inoculation trials against rinderpest carried out during that period, it is worth mentioning the work of Geert Reinders (1737-1815) in the Netherlands. He was a farmer in the Province of Groningen and a self-taught man. During his experiments Reinders noticed that calves from recovered cows were resistant to infection. This was most probably the first recognition of the phenomenon of maternally-derived immunity, since, according to him, this resistance was not of hereditary origin, depending solely on the immunity of the dam. He also noticed that the transferred protection gradually disappeared, leaving the calves just as susceptible as those from dams who had not had the disease. He also took advantage of this temporary resistance to inoculate calves with minimal risk and realised that he increased his chances of successful inoculation by repeating the procedure at different ages, because in some of the calves the first inoculation would not 'take' (i.e. have the desired effect).

Nevertheless, in the end it became obvious that inoculation was not a valid solution for rinderpest control. Not only were the losses after inoculation too high but, more importantly, the procedure perpetuated the circulation of the causative agent in the cattle population. At least all these experiments proved that smallpox was not unique in being preventable by inoculation and that the procedure, when successful, provided lifelong protection. After Jenner's discovery of vaccination against smallpox in 1796, and due to the suspected analogy between the two diseases, there were trials to vaccinate cattle against rinderpest using the smallpox vaccine.

Henri Bouley demonstrated the total lack of crossprotection between rinderpest, smallpox and vaccinia in 1865. For this purpose he sent eight cows to England, where the rinderpest epizootics were raging. These cows, which had already been used in France to produce the anti-smallpox vaccine, all contracted rinderpest. Later on, Robert Koch, working in South Africa, suggested that cattle could be protected by subcutaneous injection of blood and bile from an infected animal. This highly dangerous method was soon replaced by the use of immune serum and later by a mixture of immune serum and virulent virus.

Subsequently, the technique was improved by serial passages of the bovine virus through goats, which enabled Edwards to produce a caprinised vaccine in India in the 1920s. Trials with inactivated vaccines also took place. Finally, the successful isolation of the virus in cell culture led to the in vitro development of an attenuated strain and from this the production of a safe and highly effective vaccine.

The Plowright tissue culture vaccine has been used with great success over the past forty years to vaccinate against rinderpest and has been the major reason behind the success of the global campaigns to eradicate the disease. We may justifiably hope that rinderpest will follow smallpox into oblivion, as only the second great plague to be eliminated on earth.

Bovine Contagious Pleuropneumonia

Another important disease of cattle, bovine contagious pleuropneumonia (CBPP), also played a key role in the history of veterinary vaccinology. The disease was a real scourge in Europe during the 19th Century, reaching Belgium in 1828, the Netherlands in 1833, and the United Kingdom (UK) in 1841. Louis Willems, a Belgian physician, tried to use inoculation to prevent the disease. Willems was the son of a distiller in Hasselt; his father had a huge fattening cowshed where he held cows coming from many different herds, which provided the ideal conditions for the transmission of a contagious disease such as bovine pleuropneumonia. Willems chose to inoculate cattle at the tail, provoking large abscesses; the animals presented general clinical signs but not the typical signs of the disease (pleuropneumonia) and became protected when exposed again.

It must be noted that a procedure similar to that proposed by Willems was empirically developed in Western Africa, where cattle were vaccinated with virulent pleuropneumonia tissues; vaccination sometimes provoked exostosis leading to a horny protrusion on the nasal bone. The skulls of such vaccinated animals even led to false identification of a new species named Bos triceros.

Bovine contagious pleuropneumonia also led to the discovery of Mycoplasma by Nocard and his colleagues in 1898; for a long time Mycoplasma were called pleuropneumonia-like organisms (PPLO), a term which was applied to the group of microorganisms similar to Mycoplasma mycoides, the cause of pleuropneumonia in cattle.

Foot and Mouth Disease

The protection of herds against the consequences of FMD has been a concern for cattle breeders for centuries, probably since antiquity. Vaccination is a recent development (between the two World Wars) in the history of farm animal breeding, and was preceded by various alternative measures, all of them oriented to protect the herd from losses induced by the threatened disease. The oldest known strategy used by cattle breeders in the distant past to confer active protection on their herd was to practise 'aphtisation' as soon as the first case of FMD was observed in the herd or in the neighbourhood. The simultaneous inoculation of all animals in the herd by rubbing muzzle or lips with virulent saliva taken from the lesions of FMD-ill animals conferred a very early, strong and long lasting post-infectious immunity.

The clinical disease triggered was comparable to the spontaneous disease, but nevertheless, there were some positive aspects like the brevity of the clinical signs, the synchronisation of infection in the full herd, the absence of aggravation of virulence by passages and finally the aim of the operation, the immunity (monovalent) conferred for several years. This kind of general method for 'prevention' is well documented as having been carried out in both animal species and humans (small pox) for centuries in Asia, Africa and Europe.

The next step just before vaccine use, was the injection of therapeutic immune serums for preventing or curing FMD symptoms in cattle. Friedrich Löffler, the co-discoverer of the filterable nature of the FMD agent (1897), pioneered this new preventive means to protect herds, which was then further developed by many other researchers. After the First World War, the production of cattle immune serum was organised at industrial level in many European countries, for example, records show that nearly 13,000 cattle were treated in one year in France and that 112,000 litres of immune serum were used in nine years in Denmark in the 1920s. It is interesting to note that several authors promoted the use of immune serum associated with aphtisation to minimise the consequences of the inoculated disease.

The history of vaccination as a whole is very interlinked with the history of FMD vaccines, the progress in industrial vaccine technology offering new opportunities to modify the opinion on vaccination or on the way to use it. For this reason, the history of the development of FMD vaccines will be presented here in some detail.

The first published attempt at using a protective FMD vaccine was that of French researchers Vallée, Carré and Rinjard in 1926. Since 1922 they had been testing the action of formaldehyde on different agents of infectious diseases. In 1925 they published an article on the first vaccine, which was made from ground mucosal FMD lesions in saline buffer that had been filtered and inactivated at 20°C for 4 to 7 days with formaldehyde at 0.5%. The protection given was irregular but when present was reported as good for the standards of that time.

In 1932 in Denmark, Schmidt completed the laboratory process by the simultaneous use of aluminium hydroxide gel, a compound that had been used with formaldehyde in the domain of tetanus and diphtheria toxins by Ramon since 1924 at the Pasteur Institute in Paris. Semi-industrial production of FMD vaccine began after the technique was further improved by the team of Professor D. Waldmann, working at the German Institute of Riems Island in the Baltic Sea. Thus, the first modern technology for turning FMD viruses into antigens for vaccines was born, and it was used with almost no modification for 50 years up until the 1970s, when attempts were made in industrial production to use other inactivants like glycidaldehyde, or aziridines. Live attenuated vaccines against FMD have not succeed, even if there has been some semi-industrial production of such vaccines at certain times (reported by Kemron in Israel in eggs, Gribanov and Onufriev in the USSR in baby rabbits and Villegas in Venezuela in eggs).

Industrial development

Once the difficult process of turning the virulent FMD viruses into safe antigens was mastered, the second difficulty to solve was to obtain enough virus material for vaccine production. Once again, it was the Riems Island research team which found a solution to this problem by developing an original technology for harvesting larger quantities of virulent material, which has been known ever since as the Waldmann's method. This method was used in Europe until the 1950s and it was still being used in South America in the 1970s. To encourage the standardisation of this method worldwide, the OIE organised an International Meeting in Bern in 1947.

According to the method, the virulent material is obtained from infected cattle which are kept in a restricted stable, inoculated at the same time at several points in the tongue, and slaughtered when the tongue lesions are at their worst. All tongues are isolated and scraped to collect lymph and epithelial lesions. The carcasses are kept in the fridge for lactic maturation for 48 hours before rejoining the commercial circuits for fresh meat. The virulent tongue lesions are ground in saline buffer, centrifuged, then diluted before the inactivation step. At the earliest stage in the development of the method, one cattle dose of monovalent vaccine was a volume of 60 ml and each cattle tongue allowed for the preparation of 40 to 50 commercial cattle doses.

One disadvantage of the Waldmann's method was the necessity to use FMD-free cattle to develop large lesions after inoculation. So, as the use of vaccination progressed across the country, fewer susceptible animals were available for the production of vaccine. The second breakthrough in FMD vaccine production was made by Professor Frenkel, a Dutch scientist from the Amsterdam Veterinary Institute. Taking advantage of the work of Maitland on tissues maintained in a special medium, Prof. Frenkel had the brilliant idea of collecting epithelia fragments taken from the tongues of healthy cattle immediately after slaughter in normal abattoirs. Maintained for 48 hours or more in an appropriate medium at 37°C under oxygen bubbling, the small pieces of epithelia (the surface areas of which were equivalent to that of a hand) were infected with a virulent seed virus.

The virus multiplied in the epithelial cells and at the end of the culture time virus was present both in the epithelia and in the maintenance medium. The process was presented as experimental at the OIE meeting in Bern in 1947, but industrial development started in 1950. The concept was revolutionary for this time because the source of raw materials (tongues) was without limit in normal abattoirs, the vaccination status of the animal had no effect on virus multiplication, and the yield of FMD virus harvested per animal was 100 times more than in the Waldmann's method (400 commercial doses).

In Chile, in 1951, Espinet discovered that saponins could be used as an effective adjuvant in the aluminium hydroxide gel, which led to the first modern vaccine available for vaccination campaigns. To meet the demand, 500 l culture tanks were used for vaccine production which induced economy of scale and made each vaccine batch bigger and each vaccine dose cheaper. And the cherry on the cake was that the vaccine prepared with bovine homologous material did not induce allergic reactions in repeated vaccination campaigns, an issue which subsequently became a huge problem with the vaccines obtained using heterologous hamster cells for virus growth. The third major technical step in the progression of FMD vaccine production was the use of cells, first in monolayer, then in suspension, to satisfy the huge demand for millions of litres of vaccine for vaccination campaigns in development in Europe or in South America.

Cells in monolayer were used on an industrial scale mainly in Italy. At first, the cells used were primary or secondary kidney cells (from calves, piglets, lambs) taken from abattoirs. Subsequently, the advantages of a clean cell line like the baby hamster kidney cell line (BHK 21) isolated by Macpherson and Stocker became apparent. But soon, the capacity of plants to produce vaccine using cell monolayer culture in roller bottles was exceeded by demand; additionally, the harvest of thousands of bottles was not without risk of bacterial contamination. Consequently, the culture of cells in suspension became the method of choice for manufacturing huge volumes of vaccines.

The credit for the development of the BHK 21 cell line in suspension must go to Capstick and Telling, who carried out their work at the Pirbright Laboratory in the UK in 1962. The main advantage of this new technology was that everything could now be done in a completly closed circuit: cell growth, the infection of cells with sterile seed virus, the clarification in line of the virus harvest, its inactivation, concentration and formulation with adjuvant, and, finally, the filling of vaccine vials. At a time in the 1970s when FMD outbreaks were rare after successful mass vaccinations, virus escapes from manufacturing plants were seen as scandalous; consequently, the new process which was safely contained in a closed circuit, itself located in an appropriate containment unit, was the beginning of real biosecurity.

The unique but huge disadvantage of this process was the presence of allergens from cell culture in the vaccine and the allergic reactions this provoked during regular vaccination campaigns. It took a decade to fine-tune purification steps so that a potent, non-allergenic vaccine could be produced in huge volumes without impairing the virus yield. After the search for new adjuvants for potent FMD vaccines for pigs by McKercher and his group at Plum Island in the USA after 1965, it became obvious in the early 1970s, that oil-adjuvanted vaccines for cattle could have a promising future in regions such as South America where cattle breeding was extensive. In that region the oil-adjuvanted vaccines were well accepted both from an immunological and a political point of view, because they offered a new approach to rectify the errors of the past in disease control.

Oil-adjuvanted vaccines administered by intra-muscular route protected cattle under a great variety of breeding conditions and appeared to provide longer-lasting immunity than the previous aqueous vaccines. Injected by vaccinators in planned programmes, oiladjuvanted vaccines proved to be more efficient than classical vaccines bought by cattle breeders to comply with legislation but rarely injected. In fact, the great successes observed in FMD control in infected areas of South America are essentially due to the intensive use of oiladjuvanted vaccines of good quality.

PATHOGENS IN VACCINE DISCOVERY

Although there are comparatively far fewer completely

sequenced genomes for bacterial pathogens of farmed livestock than there are for human pathogens, several completed sequences of these pathogens have recently become available for studies in vaccine development. Cattle pathogens such as Bacillus anthracis, Mycobacterium avium subspecies paratuberculosis, Brucella abortus and Leptospira, as well as swine and poultry pathogens including Pasteurella multocida and Bordetella avium are among the veterinary pathogens with published sequences. Thus far, very few vaccine developments flowing from this genomic information have been published owing to the recent completion of the sequencing projects. However, two veterinary vaccines, identified through genomic approaches, have shown promise and are worthy of mention.

Eimeria is a protozoan parasite that causes coccidiosis in livestock and is especially costly to the poultry industry, with estimated annual worldwide losses at US$ 800 million. Efforts to develop a vaccine against this parasite have been difficult not only because of the several thousand genes and hence potential antigens encoded by this parasite, but also because the majority of candidate molecules are immunogenic, but few of these same molecules are immunoprotective (stimulate a protective immune response).

By using linkage analysis of DNA markers combined with an understanding of the parasite infection cycle, investigators have identified four key regions within the Eimeria genome capable of stimulating protective immunity. Some progress has also been made with Brucella abortus, a facultative intracellular pathogen that causes abortions in livestock. Researchers analysed sequence data of Brucella and found the exsA gene.

Examples of genome-wide applications with human pathogens are more common and provide a pathway that studies involving veterinary pathogens will no doubt follow. For example, Group A Streptococcus (GAS) has long been known to cause a variety of human diseases that range from pharyngitis to skin invasion/infection, but the molecular basis for the phenotypic differences underlying these widely varying disease presentations were unknown until recently. By comparing complete transcriptomes of a GAS pharyngeal isolate and GAS skin isolate, investigators found 89 genes differentially expressed, 24 of which were virulence genes. These investigators followed up on this observation by completely re-sequencing the pharyngeal isolate and found a 7-bp frameshift mutation in a two-component regulator encoded by covRS.

This mutation resulted in a truncation of the CovS protein, a histidine sensor. This mutation was exclusively responsible for the diverse disease phenotypes of each GAS isolate and has resulted in solid vaccine leads and intervention strategies. Perhaps one of the best-known examples of genomic application to vaccine development is with Neisseria meningitides Group B. Investigators sequenced the complete genome of this human pathogen and recombinantly expressed 350 genes predicted to encode surface-exposed or secreted proteins.

Immunological assays quickly trimmed the list of vaccine candidates to seven, of which some were later shown to be protective in animal studies. Thus, the term 'reverse vaccinology' was coined to reflect how genomic approaches allow for the design of vaccines starting from the prediction of all antigens in silico (performed by computer simulation), independently of their abundance and without the need to grow the micro-organism in vitro.

Accordingly, the genomic approach that can most readily be applied to vaccine discovery is the creation of subunit vaccines. In most cases, the production of killed or wholecell protein preparations and attenuated live vaccine strains does not need genomic technology, but rather a protein chemist or microbiologist. However, the identification of suitable antigens is crucial to successful vaccine development based on subunit approaches. Using a combination of proteomics, genomics and bioinformatics, investigators can quickly narrow the list from thousands of genes down to a few dozen vaccine candidates.

The genomics/proteomics methods define the coding capacity, and then the bioinformatic analyses trim off pseudogenes (nonfunctional or noncoding) and sequences similar to human proteins and make predictions about secreted and surface-located proteins. Then it is back to the laboratory to recombinantly produce these proteins and determine their immunogenicity, although bioinformatics can make limited predictions along these lines as well. From these exercises emerge the short list of solid vaccine candidates to test in an animal model for protection.

Genomics also has the capability to make DNA vaccination studies much more efficient. Before the genome sequence was available for Mycobacterium avium subspecies paratuberculosis, DNA vaccination was attempted for this cattle pathogen using the expression library immunisation procedure. Random expression library immunisation was used because the genome sequence was not available at the project's inception. This random cloning method meant that the majority of clones would be in the opposite orientation relative to the coding strand or would be out of frame with the coding sequence.

Therefore, many additional clones were needed to make the library truly representative of every coding sequence in the genome. An approximate total of 16,500 clones were used to immunise mice in that study. With the genome sequence now complete, a directed expression library immunisation project, in which each clone faithfully represented a single coding sequence, could be initiated. This method has the advantage that fewer clones are needed, making resulting clone pools less complex and there is no 'garbage' or nonfunctional clones such as those in the opposite orientation or out of frame. For such a study, only 4,350 clones would be needed because that is the total number of genes present in the M. avium subspecies paratuberculosis genome. An added benefit is that fewer mice would be needed to test the clone pools.

Genomic approaches can also identify the best targets for knockout mutations that enable engineering of attenuated vaccine strains. However, as yet, there are no published studies for bacterial pathogens that demonstrate a genomewide approach that can identify a target, knockout this target, and show both attenuation and protection in an animal host. Rather, the literature reports studies in which genomics has been used to genetically define a known vaccine strain. The Salmonella typhimurium vaccine strain is protective in mice and lacks the transcriptional regulator RfaH.

Use of whole genome microarrays identified the RfaH-dependent genes giving investigators insight into the mechanism of attenuation for this vaccine strain. The most famous example of the use of genomics to define an attenuated bacterial vaccine strain is M. bovis BCG Bacillus Calmette-Guérin - named after the the French scientists Calmette and Guérin), which is the most widely used global vaccine to prevent human tuberculosis (TB). Over 3 billion individuals have been vaccinated with BCG without major side effects.

The BCG vaccine strain was derived long ago from a fully virulent isolate of M. bovis by prolonged serial passage of the bacterium resulting in its attenuation. However, the molecular basis for this attenuation was never understood until the complete genome sequences of M. tuberculosis, the causative agent of TB, and M. bovis BCG became available. Genomic comparison of these two species revealed one region of deletion in BCG, termed RD1. This region contains the well-known antigen ESAT-

6, a secreted antigenic target that strongly induces Th1 immune responses.

More than any other benefit, whole genome analysis of pathogens enables the targeted selection of protective immunogens encoded by the disease-causing pathogen. This allows investigators to move away from empirical approaches in vaccine development towards a more focused, logical development and discovery of protective DNA segments and proteins.

REFERENCES

Hendolin P, Markkanen A, Ylikoski J, Wahlfors J.J., "Use of multiplex PCR for simultaneous detection of four bacterial species in middle ear effusions", *Journal of Clinical Microbiology,* 1997.

Hugo W. B., and A. D. Russell, *Pharmaceutical microbiology,* 3rd ed Singapore: P.G. Pub., 1983

Huws, S. A., Smith, A. W. Enright, M. C., Wood, P. J. and Brown, M. R. W., "Amoebae promote persistence of epidemic strains of MRSA", *Environmental Microbiology,* 2006.

Pitkäranta A, Arruda E, Malmberg H, Hayden F.G., "Detection of rhinovirus in sinus brushings of patients with acute community-acquired sinusitis by reverse-transcription -PCR", *Journal of Clinical Microbiology,* 1997.

5

Development of Antimicrobial Drugs

Antimicrobial drugs can be used for either prophylaxis (prevention) or treatment of disease caused by bacteria, fungi, viruses, protozoa, or helminths. Antibacterial drugs are derived from bacteria or molds or from de novo synthesis. "Antibiotic," which is often used synonymously with "antibacterial drug," technically refers only to antimicrobials derived from bacteria or molds. Antibacterials have many mechanisms of action, including inhibiting cell wall synthesis, activating enzymes that destroy the cell wall, increasing cell membrane permeability, and interfering with protein synthesis and nucleic acid metabolism.

Antibacterials sometimes interact with other drugs, raising or lowering serum levels of other drugs by increasing or decreasing their metabolism or various other mechanisms. The most clinically important interactions involve drugs with a low therapeutic ratio (ie, toxic levels are close to therapeutic levels). Many antibacterials are chemically related and are thus grouped into classes. Although drugs within each class share structural and functional similarities, they often have different pharmacology and spectra of activity.

USE OF ANTIBACTERIAL DRUGS

Antibacterials should be used only if clinical or laboratory evidence suggests *bacterial* infection. Use for viral illness or undifferentiated fever is inappropriate, subjects the patient to drug complications without any benefit, and contributes to bacterial resistance. Certain bacterial infections (eg, abscesses, infections with foreign bodies) require surgical intervention and do not respond to antibiotics alone.

Cultures and antibiotic sensitivities are essential for selecting a drug for serious infections. However, treatment often must begin before culture results are available, necessitating selection according to the most likely infecting organisms (empiric antibiotic selection). Whether chosen according to culture results or empirically, drugs used should possess the narrowest spectrum of activity that will control the infection.

For empiric treatment of serious infections that may involve any one of several pathogens (eg, fever in a neutropenic patient) or that may be due to multiple pathogens (eg, polymicrobial anaerobic infection), a broad spectrum of activity is desirable. The most likely organisms and the organisms' susceptibility to antibacterials vary according to geography (within cities or even within a hospital) and can change from month to month. Bactericidal drugs kill bacteria in vitro. Bacteriostatic drugs slow or stop in vitro bacterial growth but depend on body defenses to kill bacteria.

Quantitative methods identify the minimum in vitro concentration at which an antibiotic can inhibit growth (minimum inhibitory concentration, or MIC) or kill (minimum bactericidal concentration, or MBC). However, in vivo antibacterial effectiveness involves other factors, including pharmacology (eg, absorption, distribution, concentration in fluids and tissues, protein binding, and rate of excretion or metabolism), the presence of drug interactions or inhibiting substances, and host defense mechanisms. Usually, greater in vitro killing power is important only if local or systemic host defenses are weak (eg, in endocarditis, meningitis, serious infections in neutropenic or other immunocompromised patients).

The predominant determinant of bacteriologic response to antibiotics is either the duration that blood levels of the antibiotic exceed the MIC (time-dependence) or the peak blood level relative to MIC (concentration-dependence). The β-lactams and vancomycin exhibit time-dependent bactericidal activity. Increasing their concentrations above the MIC does not increase their rate of bactericidal activity, and their in vivo killing is generally slow. In addition, because there is either no or very brief residual inhibition of bacterial growth after concentrations fall below the MIC (postantibiotic effect, or PAE), β-lactams and vancomycin are most often effective when serum levels of free drug (ie, drug not bound to serum protein) exceed the MIC for = 50% of the time.

The long serum half-life of ceftriaxone allows free serum levels to exceed the MIC of very susceptible pathogens for the entire 24-h dosing interval. However, frequent dosing or continuous infusion is required for other β-lactams that have serum half-lives of = 2 h. For vancomycin, trough levels should be maintained at 10 to 15 μg/mL.

Aminoglycosides, fluoroquinolones, and daptomycin exhibit concentration-dependent bactericidal activity. Increasing their concentrations from levels slightly above the MIC to levels far above the MIC increases their rate of bactericidal activity and decreases the bacterial load. In addition, after brief exposure to concentrations above the MIC, aminoglycosides and fluoroquinolones exhibit a PAE on residual bacteria, the duration of which is also concentration-dependent. If PAEs are long, drug levels can be below the MIC for extended periods without loss of efficacy, allowing less frequent dosing. Consequently, aminoglycosides and fluoroquinolones are usually most effective as intermittent boluses that achieve peak free serum levels = 10 times the MIC of the infecting organism.

Combinations of antibiotics are often necessary in serious infections, either because they provide treatment for multiple possible species of infecting bacteria or because they act synergistically against a single species of bacteria. Synergism is usually defined as more rapid and complete bactericidal action from a combination of antibiotics (usually a cell wall–active agent, ie, a β-lactam or vancomycin, plus an aminoglycoside) than could be achieved by either antibiotic alone.

Oral administration provides excellent blood levels of many antibiotics and does so nearly as rapidly as IV administration. IV administration is preferred if oral antibiotics cannot be tolerated (eg, due to vomiting) or absorbed (eg, due to malabsorption after intestinal surgery); intestinal motility is impaired (eg, due to opioids); no oral preparation is available (eg, aminoglycosides); or a patient is critically ill, in which case GI tract perfusion may be impaired or even the brief delay with oral administration may be detrimental.

Antibiotics should be continued until objective evidence of systemic infection (eg, absence of fever, symptoms, and abnormal laboratory findings) is absent for several days. Courses of therapy for some infections (eg, endocarditis, TB, osteomyelitis) are weeks or months long to prevent relapse.

Doses and scheduling of antibacterials may need to be adjusted in infants, the elderly, and patients with renal failure. Commonly used antibacterials that need dose adjustment in hepatic insufficiency include cefoperazone, ceftriaxone, chloramphenicol, clindamycin, metronidazole, nafcillin, and rifampin. Penicillins, cephalosporins, and erythromycin are among the safest antibacterials during pregnancy; tetracyclines are contraindicated. Most antibiotics reach sufficient concentrations in breast milk to affect a breastfed baby, sometimes contraindicating their use in nursing women.

Complications of antibiotic therapy include superinfection by nonsusceptible bacteria or fungi and, commonly, cutaneous, renal, hematologic, and GI adverse effects. Adverse effects frequently require stopping the offending drug and substituting another antibiotic to which the pathogen is susceptible; sometimes, no alternatives exist.

Antibiotic Resistance

Resistance to an antibiotic may be inherent in a particular bacterial species or may be acquired as a result of mutations or acquisition of genes from another organism that encode for antibiotic resistance. Resistance genes can be transmitted between 2 bacterial cells by transformation, transduction (infection by a bacteriophage), or conjugation (exchange of genetic material in the form of either plasmids, which are pieces of independently replicating extrachromosomal DNA, or transposons, which are movable pieces of chromosomal DNA). Plasmids and transposons can rapidly disseminate resistance genes.

Antibiotic use preferentially eliminates nonresistant bacteria, increasing the proportion of resistant bacteria that remain. This is true not only for pathogenic bacteria but also for normal flora; resistant normal flora serves as a reservoir for resistance genes that can spread to future pathogens.

SOME ANTIBACTERIAL DRUGS

Aminoglycosides

Aminoglycosides are a group of antibiotics that are used to treat certain bacterial infections. This group of antibiotics includes at least eight drugs: amikacin, gentamicin, kanamycin, neomycin, netilmicin, paromomycin, streptomycin, and tobramycin. All of these drugs have the same basic chemical structure.

Aminoglycosides are primarily used to combat infections due to aerobic, Gram-negative bacteria. These bacteria can be identified by their reaction to Gram's stain. In Gram's staining, a film of material containing the possible bacteria is placed on a glass slide and dried. The slide is stained with crystal violet for one minute, cleaned off with water and then placed into a solution of Gram's iodine solution for one minute. The iodine solution is rinsed off and the slide is immersed in 95% ethyl alcohol. The slide is then stained again with reddish carbolfuchsin or safranine for 30 seconds, rinsed in water, dried and examined.

Gram-positive bacteria retain the violet purple stain. Gram-negative bacteria accept the red stain. Bacteria that can successfully be combated with aminoglycosides include *Pseudomonas*, *Acinetobacter*, and *Enterobacter* species, among others. Aminoglycosides are also effective against mycobacteria, the bacteria responsible for tuberculosis.

The aminoglycosides can be used against certain Gram-positive bacteria, but are not typically employed because other antibiotics are more effective and have fewer side effects. Aminoglycosides are ineffective against anaerobic bacteria (bacteria that cannot grow in the presence of oxygen), viruses, and fungi. And only one aminoglycoside, paromomycin, is used against parasitic infection. Like all other antibiotics, aminoglycosides are not effective against influenza, the common cold, or other viral infections.

Precautions: Pre-existing medical conditions—such as kidney disease, eighth cranial nerve disease, myasthenia gravis, and Parkinson's disease—should be discussed prior to taking any aminoglycosides. Pregnant women are usually advised against taking aminoglycosides, because their infants may suffer damage to their hearing, kidneys, or sense of balance. However, those factors need to be considered alongside the threat to the mother's health and life in cases of serious infection. Aminoglycosides do not pass into breast milk to any great extent, so nursing mothers may be prescribed aminoglycosides without injuring their infants.

Description: Streptomycin, the first aminoglycoside, was isolated from *Streptomyces griseus* in the mid-1940s. This antibiotic was very effective against tuberculosis. One of the main drawbacks to streptomycin is its toxicity, especially to cells in the inner and middle ear and the kidney. Furthermore, some strains of tuberculosis are resistant to treatment with streptomycin. Therefore, medical researchers have put considerable effort into identifying other antibiotics with streptomycin's efficacy, but without its toxicity.

Aminoglycosides are absorbed very poorly from the gastrointestinal tract; in fact, aminoglycosides taken orally are excreted virtually unchanged and undiminished in quantity. The route of drug administration depends on the type and location of the infection being treated. The typical routes of administration are by intramuscular (injection into a muscle) or intravenous injection (injection into a vein), irrigation, topical skin application, or inhalation. If the infection being treated involves the central nervous system, the drug can be injected into the spinal canal.

The bactericidal ability of aminoglycosides has not been fully explained. It is known that the drug attaches to a bacterial cell wall and is drawn into the cell via channels made up of the protein, porin. Once inside the cell, the aminoglycoside attaches to the cell's ribosomes. Ribosomes are the intracellular structures responsible for manufacturing proteins. This attachment either shuts down protein production or causes the cell to produce abnormal, ineffective proteins. The bacterial cell cannot survive with this impediment.

Antibiotic treatment using aminoglycosides may pair the drug with a second type of antibiotic, usually a beta-lactam or vancomycin, administered separately. Beta-lactams disrupt the integrity of the bacteria cell wall, making it more porous. The increased porosity allows more of the aminoglycoside into the bacteria cell.

Traditionally, aminoglycosides were administered at even doses given throughout the day. It was thought that a steady plasma concentration was necessary to combat infection. However, this administration schedule is time and labor intensive. Furthermore, administering a single daily dose can be as effective, or more effective, than several doses throughout the day.

Dosage depends on the patient's age, weight, gender, and general health. Since the drug is cleared by the kidneys, it is important to assess any underlying problems with kidney function. Kidney function is assessed by measuring the blood levels of creatinine, a protein normally found in the body. If these levels are high, it is an indication that the kidneys may not be functioning at an optimal rate and dosage will be lowered accordingly.

Risks: Aminoglycosides have been shown to be toxic to certain cells in the ears and in the kidneys. Approximately 5-10% of the people who are treated with aminoglycosides experience some side effect, affecting their hearing, sense of balance, or kidneys. However, in most cases the damage is minor and reversible once medication is stopped.

If cells in the inner ear are damaged or destroyed, an individual may experience a loss of balance and feelings of dizziness. Damage to the middle ear may result in hearing loss or tinnitus. Neomycin, kanamycin, and amikacin are the most likely to cause problems with hearing, and streptomycin and gentamicin carry the greatest risk of causing vertigo and loss of balance. Kidney damage, apparent with changes in urination frequency or urine production, is most likely precipitated by neomycin, tobramycin, and gentamicin.

Young children and the elderly are at the greatest risk of suffering side effects. Excessive dosage or poor clearance of the drug from the body can be injurious at any age.

Less common side effects include skin rashes and itching. Very rarely, certain aminoglycosides may cause difficulty in breathing, weakness, or drowsiness. Gentamicin, when injected, may cause leg cramps, skin rash, fever, or seizures.

If side effects linger or become worse after medication is stopped, it is advisable to seek medical advice. Side effects that may be of concern include tinnitus or loss of hearing, dizziness or loss of balance, changes in urination frequency or urine production, increased thirst, appetite loss, and nausea or vomiting.

Normal results: At the proper dosage and in the presence of gram-negative enteric (intestinal) bacteria, aminoglycosides are very effective in treating an infection.

Abnormal results: In some cases, bacteria are resistant to antibiotics that would normally kill them. This resistance becomes apparent after repeated exposure to the antibiotic and arises from a mutation that alters the bacteria's susceptibility to the drug. Various degrees of resistance have been observed in bacteria that normally would be destroyed by aminoglycosides. In general, though, aminoglycoside effectiveness has held up well over time.

Chloramphenicol

Chloramphenicol represents the product of years of antibiotic development. Due to its pH, it shines above most other antibiotics in terms of ability to penetrate. Chloramphenicol can easily pass deeply through purulent material to the organisms hiding within, through cell membranes to attack parasites living within, and into organs where other antibiotics cannot go.

It acts on the protein manufacturing system of bacteria (the intracellular ribosomes) yet does not effect mammalian, reptilian, or avian ribosomes. Highly susceptible bacteria are killed while others are merely rendered unable to divide. Chloramphenicol has an especially broad spectrum of activity against numerous aerobic bacteria, mycoplasma, Chlamydial organisms, and anaerobic bacteria.

Chloramphenicol may be given orally or topically, usually three times daily. Peak activity occurs approximately 30 minutes after an oral dose except in the nervous system where several hours are required for penetration of the "blood/brain barrier." This medication is an especially good choice for infections where:

— There is a necrotic or walled off area with inner infection (pneumonia is a good example)
— The central nervous system or eye is involved
— The prostate gland is involved
— Intracellular parasites are involved (Chlamydia, mycoplasma, rickettsia)

In some individuals, chlorphenicol use may induce what are called "blood dyscrasias." This means that abnormal blood cells can be produced or that production of normal cells can be halted due to an action of chloramphicol on patient bone marrow. This reaction is especially of concern in cats and chloramphenicol should be used with caution in this species (special dosing, perhaps even some blood monitoring).

Chloramphenicol use may accumulate to toxic levels in very young animals (in first few weeks of life) and as they are not able to remove it from their bodies as effectively as adult animals. For this reason, it is best not to give this medication during pregnancy or lactation.

Nausea, diarrhea, and appetite loss are relatively common (and usually minor) side effects of this medication but if they occur, another medication can be selected.

Daptomycin

Daptomycin is a novel lipopeptide antibiotic used in the treatment of certain infections caused by Gram-positive organisms. It is a naturally-occurring compound found in the soil saprotroph *Streptomyces roseosporus*. Its distinct mechanism of action means that it may be useful in treating infections caused by multi-resistant bacteria. It is marketed in the United States under the trade name Cubicin.

Daptomycin has a distinct mechanism of action, disrupting multiple aspects of bacterial cell membrane function. It appears to bind to the membrane and cause rapid depolarisation, resulting in a loss of membrane potential leading to inhibition of protein, DNA and RNA synthesis, which results in bacterial cell death.

The bactericidal activity of daptomycin is concentration-dependent. There is *in vitro* evidence of synergy with β-lactam antibiotics.

Daptomycin is active against Gram-positive bacteria only. It has proven *in vitro* activity against enterococci (including glycopeptide-resistant Enterococci (GRE)), staphylococci (including methicillin-resistant *Staphylococcus aureus*), streptococci and corynebacteria.

Clinical use: Daptomycin is approved in the United States for skin and skin structure infections caused by Gram-positive infections, *Staphylococcus aureus* bacteraemia and right-sided *S. aureus* endocarditis.

Daptomycin has been shown to be not inferior to standard therapies (nafcillin, oxacillin, flucloxacillin or vancomycin) in the treatment of bacteraemia and right-sided endocarditis caused by *Staphylococcus aureus*. In Phase III clinical trials, limited data showed that daptomycin was associated with poor outcomes in patients with left-sided endocarditis. It is inactivated by pulmonary surfactants and is not indicated for the treatment of pneumonia.

Daptomycin has not been studied in patients with prosthetic valve endocarditis or meningitis.

Dosage and presentation: In skin and soft tissue infections, 4 mg/kg daptomycin is given intravenously once daily. For *S. aureus* bacteraemia or right-sided endocarditis, the approved dose is 6 mg/kg given intravenously once daily. The dose of daptomycin must be reduced in renal impairment. There is no information available on dosing in people less than 18 years of age.

Daptomycin is supplied as a sterile preservative-free pale yellow to light brown lyophilised 500 mg cake that must be reconstituted with 0.9% saline prior to use.

Fluoroquinolones

Fluoroquinolones are medicines that kill bacteria or prevent their growth. Fluoroquinolones are antimicrobials, medicines used to treat infections caused by microorganisms. Physicians prescribe these drugs for bacterial infections in many parts of the body. For example, they are used to treat bone and joint infections, skin infections, urinary tract infections, inflammation of the prostate, serious ear infections, bronchitis, pneumonia, tuberculosis, some sexually transmitted diseases (STDs), and some infections that affect people with AIDS.

Fluoroquinolones are available only with a physician's prescription and are sold in tablet and injectable forms. Examples of these medicines are moxifloxacin (Avelox), ciprofloxacin (Cipro), ofloxacin (Floxin), levofloxacin (Levaquin), lomefloxacin (Maxaquin), norfloxacin (Noroxin), enoxacin (Penetrex), gatifloxacin (Tequin), and sparfloxacin (Zagam).

In the wake of the anthrax terrorist attacks in the United States in 2001, ciprofloxacin received extensive media attention because it was the only drug labeled as approved by the Food and Drug Administration (FDA) for both prophylaxis and treatment of inhalation anthrax (the most serious form of the disease).

Dosage: The recommended dosage depends on the type and strength of fluoroquinolone, and the kind of infection for which it is being taken. Check with the physician who prescribed the drug or the pharmacist who filled the prescription for the correct dosage. To make sure the infection clears up completely, take the medicine for as long as it has been prescribed. Do not stop taking the drug just because symptoms begin to improve. Symptoms may return if the drug is stopped too soon.

Fluoroquinolones work best when they are at constant levels in the blood. To help keep levels constant, take the medicine in doses spaced evenly through the day and night. Do not miss any doses. For best results, take this medicine with a full glass of water and drink several more glasses throughout the day, every day during treatment with the drug. The extra water will help prevent some side effects. Some fluoroquinolones should be taken on an empty stomach; others may be taken with meals.

Precautions: An important precaution for any antibiotic is that unnecessary use or abuse of antibiotics can encourage drug-resistant strains of bacteria to develop and proliferate. These drug-resistant strains then become difficult, or even impossible, to treat. Bacteria found in hospitals appear to have become especially resilient, and are causing increasing difficulty for patients and the doctors treating them. Following the U.S. 2001 anthrax attacks, for example, the American Medical Association urged its members not to prescribe ciprofloxacin unnecessarily. One fear is that the overuse of the drug could reduce its effectiveness against infections such as typhoid fever, hospital-acquired pneumonia, and others.

Research suggests that fluoroquinolones may cause bone development problems in children and teenagers. Infants, children, teenagers, pregnant women, and women who are breastfeeding should not take this medicine unless directed to do so by a physician.

Although such side effects are rare, some people have had severe and life-threatening reactions to fluoroquinolones. Call a physician immediately if any of these signs of a dangerous reaction occur:

— swelling of the face and throat
— swallowing problems
— shortness of breath
— rapid heartbeat
— tingling of fingers or toes
— itching or hives
— loss of consciousness

Some fluoroquinolones may weaken the tendons in the shoulder, hand, or heel, making the tendons more likely to tear. Anyone who notices pain or inflammation in these or other tendon areas should stop taking the medicine immediately and call a physician. Rest and avoid exercise until the physician determines whether the tendons are damaged. If the tendons are torn, surgery may be necessary to repair them.

These medicines make some people feel drowsy, dizzy, lightheaded, or less alert. Anyone who takes these drugs should not drive, use machines or do anything else that might be dangerous until they have found out how the drugs affect them.

This medicine may increase sensitivity to sunlight. Even brief exposure to sun can cause a severe sunburn or a rash. While being treated with fluoroquinolones, avoid being in direct sunlight, especially between 10 a.m. and 3 p.m.; wear a hat and tightly woven clothing that covers the arms and legs; use a sunscreen with a skin protection factor (SPF) of at least 15; protect the lips with a sun block lipstick; and do not use tanning beds, tanning booths, or sunlamps.

Do not take antacids that contain aluminum, calcium, or magnesium at the same time as fluoroquinolones. The antacids may keep the fluoroquinolones from working as they should. If antacids are needed, take them at least two hours before or two hours after taking norfloxacin or ofloxacin, at least four hours before or two hours after taking ciprofloxacin. Follow the same instructions for taking sucralfate (Carafate), a medicine used to treat stomach ulcers and other irritation in the digestive tract and mouth.

Anyone who has had unusual reactions to fluoroquinolones or related medicines such as cinoxacin (Cinobac) or nalidixic acid (NegGram) in the past should let his or her physician know before taking the drugs again. The physician should also be told about any allergies to foods, dyes, preservatives, or other substances. Before using fluoroquinolones, people with any of these medical problems should make sure their physicians are aware of their conditions:

— kidney disease
— liver disease with kidney disease
— diseases of the brain or spinal cord, including hardening of the arteries in the brain, epilepsy, and other seizure disorders

Taking fluoroquinolones with certain other drugs may affect the way the drugs work or may increase the chance of side effects.

Lincosamides

Lincosamides (eg. lincomycin, clindamycin) are a class of drugs which bind to the 23s portion of the 50S subunit of bacterial ribosomes and inhibit early elongation of peptide chain by inhibiting transpeptidase reaction. In this sense they have a similar action to macrolides.

Lincomycin has been superseded by clindamycin, which exhibits improved antibacterial activity. Clindamycin also exhibits some activity against parasitic protozoa, and has been used in toxoplasmosis and malaria.

They are normally used to treat staphylococci and streptococci, and have proved useful in treating *Bacteroides fragili* and some other anaerobes.

Target bacteria may alter the drug's binding site (similar to resistance found in macrolides and streptogramins). The resistance mechanism is methylation of the 23s binding site. If this occurs then the bacteria are resistant to both the macrolides and the lincosamides.

The lincosamides, as the hydrochloride salt, are bitter to taste, so for oral formulation they are given as the palmitate esters, or formulated in capsules. Clindamycin is given intravenously as clindamycin phosphate, which is then converted into active clindamycin wihin the body.

Oxazolidinones

Oxazolidinones are a class of compounds containing 2-oxazolidone in the structure. In chemistry, they are useful as Evans auxiliaries, which are used for chiral synthesis. Usually, the acid chloride substrate reacts with the oxazolidinone to form an imide. Substituents at the 4 and 5 position of the oxazolidinone direct any aldol reaction to the alpha position of the carbonyl of the substrate.

Oxazolidinones are used as antibiotics. Some of the most important oxazolidinones are the last generation of antibiotics used against gram-positive pathogens, including superbugs such as *Methicillin-resistant Staphylococcus aureus*. Developed during the nineties when several bacterial strains were becoming resistant against such antibiotics as vancomycin. These antibiotics are considered as a choice of last resort where every other antibiotic therapy has failed. Linezolid (Zyvox), the only currently available agent in this class, is available for intravenous administration and also has the advantage of having excellent oral bioavailability.

The first commercially available oxazolidinone antibiotic was Linezolid. In 2002 AstraZeneca introduced the AZD2563 antibiotic. Results indicate that AZD2563 has excellent, targeted activity against all common gram-positive bacteria, regardless of resistance to other classes of antibiotics. The anticoagulant rivaroxaban, an oxazolidinone derivative, is undergoing trials.

Streptogramin

Streptogramin antibiotics represent a unique class of antibacterials in the each member of the class consists of at least 2 structurally unrelated molecules: group a streptogramins (macrolactones) and group B streptogramins (cyclic hexadepsipeptides). Both group A and group B streptogramins inhibit protein synthesis at the ribosomal level, and they act synergistically against many isolates their combination generating bactericidal activities and reducing the possibility of emergencies of resistant strains.

The mechanisms of acquired resistance to group B streptogramins remain unaffected by target modifications and active efflux. The pharmacokinetic parameters of group A and group B streptogramins in blood are quite similar. In addition, both the A and B group penetrate and accumulate in macrophages and in the bacterial gegetations of experimental endocarditis. Until recently, the complex and irregular composition of naturally occurring pristinamycin and virginiamycin, as well as the unavailability of soluble forms, have limited the clinical development of streptogramins.

The synthesis of water soluble derivatives of pristinamycin IA and IIB has now allowed the development of injectable streptogramins with fixed compositions. This unique class of antibacterials will have a significant clinical impact in a world of increasing multidrug resistance affecting the Gram-positive cocci, especially staphylococci and pneumococci. The absence of cross-resistance to macrolides in many of these isolates and the rapid antibacterial killing against these species bright future for this class of antibiotics.

Macrolide

The macrolides are a group of drugs (typically antibiotics) whose activity stems from the presence of a *macrolide ring*, a large macrocyclic lactone ring to which one or more deoxy sugars, usually cladinose and desosamine, may be attached. The lactone rings are usually 14, 15 or 16-membered. Macrolides belong to the polyketide class of natural products.

Antibiotic macrolides are used to treat infections such as respiratory tract and soft tissue infections. The antimicrobial spectrum of macrolides is slightly wider than that of penicillin, and therefore macrolides are a common substitute for patients with a penicillin allergy. Beta-hemolytic streptococci, pneumococci, staphylococci and enterococci are usually susceptible to macrolides. Unlike penicillin, macrolides have been shown to be effective against mycoplasma, mycobacteria, some rickettsia and chlamydia.

The mechanism of action of the macrolides is inhibition of bacterial protein biosynthesis by binding reversibly to the subunit 50S of the bacterial ribosome, thereby inhibiting translocation of peptidyl tRNA. This action is mainly bacteriostatic, but can also be bactericidal in high concentrations. Macrolides tend to accumulate within leukocytes, and are therefore actually transported into the site of infection.

A novel, non-antibiotic, anti-inflammatory effect of 14-membered macrolides was discovered in Japan, which is especially effective in improving control of diffuse panbronchiolitis(DPB). Research on the anti-inflammatory properties of the macrolide ring is ongoing.

The primary means of bacterial resistance to macrolides occurs by post-transcriptional methylation of the 23S bacterial ribosomal RNA. This acquired resistance can be either plasmid-mediated or chromosomal, i.e through mutation, and results in cross-resistance to macrolides, lincosamides, and streptogramins (an MLS-resistant phenotype). Two other types of acquired resistance rarely seen include the production of drug-inactivating enzymes (esterases or kinases) as well as the production of active ATP-dependent efflux proteins that transport the drug outside of the cell.

ANTIVIRAL DRUGS

The emergence of antivirals is the product of a greatly expanded knowledge of the genetic and molecular function of organisms, allowing biomedical researchers to understand the structure and function of viruses, major advances in the techniques for finding new drugs, and the intense pressure placed on the medical profession to deal with the human immunodeficiency virus (HIV), the cause of the deadly acquired immunodeficiency syndrome (AIDS) pandemic.

Almost all anti-microbials, including anti-virals, are subject to drug resistance as the pathogens evolve to survive exposure to the treatment. As of 2007, only smallpox has been successfully eradicated, and Poliomyelitis eradication is still underway. Both of these efforts are using vaccines.

Virus Life Cycle

Viruses consist of a genome and sometimes a few enzymes stored in a capsule made of protein, and sometimes covered with a lipid layer. Viruses cannot reproduce on their own, so they propagate by subjugating a host cell to produce copies of themselves, thus producing the next generation. Researchers working on such "rational drug design" strategies for developing antivirals have tried to attack viruses at every stage of their life cycles. Viral life cycles vary in their precise details depending on the species of virus, but they all share a general pattern:

— Attachment to a host cell.
— Release of viral genes and possibly enzymes into the host cell.
— Replication of viral components using host-cell machinery.
— Assembly of viral components into complete viral particles.
— Release of viral particles to infect new host cells.

Inadequacy of Vaccines

Vaccines attack viruses when they are in the "complete particle" stage, outside of the organism's cells. They traditionally consist of a weakened or killed version of a pathogen, though more recently "subunit" vaccines have been devised that consist strictly of protein targets from the pathogen. They stimulate the immune system without doing serious harm to the host, and so when the real pathogen attacks the subject, the immune system responds to it quickly and blocks it.

Vaccines are very effective on stable viruses, but are of limited use in treating a patient who has already been infected. They are also difficult to successfully deploy against rapidly mutating viruses, such as influenza (the vaccine for which is updated every year) and HIV. These two gaps are where antiviral drugs become useful.

Anti-viral Targeting Technique

The general idea behind modern antiviral drug design is to

identify viral proteins, or parts of proteins, that can be disabled. These "targets" should generally be as unlike any proteins or parts of proteins in humans as possible, to reduce the likelihood of side effects. The targets should also be common across many strains of a virus, or even among different species of virus in the same family, so a single drug will have broad effectiveness. For example, a researcher might target a critical enzyme synthesized by the virus, but not the patient, that is common across strains, and see what can be done to interfere with its operation.

Once targets are identified, candidate drugs can be selected, either from drugs already known to have appropriate effects, or by actually designing the candidate at the molecular level with a computer-aided design program. The target proteins can be manufactured in the lab for testing with candidate treatments by inserting the gene that synthesizes the target protein into bacteria or other kinds of cells. The cells are then cultured for mass production of the protein, which can then be exposed to various treatment candidates and evaluated with "rapid screening" technologies.

Antiviral Active Substances

There are three types of antiviral active substances which inhibit the proliferation of influenza viruses at various phases: the two neuraminidase inhibitors zanamivir and oseltamivir; amantadine and rimantadine; and finally ribavirin. However, the latter drug is not used to treat flu even though inhibition of viral proliferation has been observed in vitro.

Neuraminidase Inhibitors

The two neuraminidase inhibitors approved for use in Switzerland - oseltamivir (Tamiflu®) and zanamivir (Relenza®) - have a comparable action against influenza A and influenza B viruses. Both can be used for pre- and post-exposure prophylaxis. In order to derive the greatest benefit from these drugs, they must be administered as early as possible, preferably within 36 hours of the first symptoms appearing.

Neuraminidase inhibitors block the viral enzyme neuraminidase, thus preventing the decomposition of viral receptors on the surface of the host cell. As a result, newly formed viruses from the host cell stay attached to the receptors and cannot spread further. This reduces the quantity of viruses that are released and are capable of infecting other cells.

Neuraminidase inhibitors reduce both the duration of influenza and its intensity and reduce viral secretion. The rate of complications is also lower in patients given neuraminidase inhibitors. Since neuraminidase inhibitors do not affect the immune response to vaccination against flu, they can also be used prophylactically following flu vaccination to protect the individual against infection with influenza until the immunity induced by vaccination has developed.

Although experience to date has shown that influenza A viruses develop less resistance to neuraminidase inhibitors than to amantadine, there have been isolated cases of viruses resistant to oseltamivir. Studies of children with seasonal influenza in Japan have also shown that resistant pathogens develop more rapidly in children than in adults. Further studies are needed to clarify the risk of resistance developing - the theoretical possibility of cross-resistance developing to zanamivir should also be borne in mind.

Oseltamivir

Oseltamivir is a product which is transformed into the active metabolite oseltamivir carboxylate in the epithelium of the gut and/or in the liver. There are two oral forms of oseltamivir on the market: capsules (10 capsules of Tamiflu® each containing 75 mg) for adults and children over 13 and a paediatric suspension (Tamiflu® suspension 52 ml containing 12 mg/ml) for children from the age of 1. Nausea and vomiting are the most common adverse effects compared to placebo. These effects are transient and occur after the first dose. There are currently no data demonstrating the efficacy of Tamiflu® in individuals with asthma, chronic respiratory diseases, immunosuppression or other chronic conditions.

Zanamivir

Zanamivir (Relenza®) is supplied for (oral) inhalation in a disc inhaler (disc containing 20 or 28 doses of 5 mg zanamivir per single dose). Zanamivir is approved for the treatment of children over 5 and for prophylaxis in children over 12. The adverse effects associated with zanamivir are comparable to those associated with placebo. To date there have been no studies published on the use of Relenza® in patients with severe asthma, severe chronic respiratory diseases, immunosuppression or severe chronic disorders, or on its therapeutic use in children under 5 or prophylactic use in children under 12.

The benefits and risks must be weighed carefully before giving the drug to patients with asthma or chronic obstructive pulmonary disease (COPD). In particular, patients should be informed about the risk of bronchospasm while using zanamivir and provided with a fast-acting bronchodilator.

Amantadine and rimantadine

Amantadine and rimantadine are effective against influenza A viruses and have the same mode of action. Only amantadine hydrochloride (Symmetrel®) is approved for use in Switzerland. Amantadine and rimantadine act on the M2 protein of influenza A viruses. Inhibition of the M2 protein prevents the viral genome inside the cell from being released once the virus has penetrated the cell. Several studies have shown that prophylactic use of drugs containing amantadine can reduce the attack rate of influenza A viruses by 70 to 90%.

In addition, administration of amantadine during the first 48 hours after flu symptoms become manifest can considerably reduce the intensity of the symptoms and the duration of the fever. However, it is not known whether this therapy is also capable of reducing the rate of complications following influenza A infection in individuals with risk factors.

The adverse effects of amantadine relate to the central nervous system and the gastrointestinal tract. They disappear within 1 or 2 days following discontinuation of the medication. Antiviral therapy with amantadine and rimantadine can lead to the development of drug-resistant influenza A viruses. The secretion of resistant viruses can be observed in treated individuals within 2 to 5 days of the start of therapy.

REFERENCES

Levy S.B (ed), "Drug Resistance: The New Apocalypse", *Trends Microbiol*, 1994.

Smith, A. W., "Biofilms and Antibiotic Therapy: Is there a Role for Combating Bacterial Resistance by the Use of Novel Drug Delivery Systems?", *Advanced Drug Delivery Reviews*, 2005.

6

Advanced Food Microbiology

The field of food microbiology is a very broad one, encompassing the study of microorganisms which have both beneficial and deleterious effects on the quality and safety of raw and processed foods.

Food scientists are generally not directly involved with the creation of genetically modified (bio-engineered) foods. The primary tool of microbiologists is the ability to identify and quantitate food-borne microorganisms; however, the inherent inaccuracies in enumeration processess, and the natural variation found in all bacterial populations complicate the microbiologists job.

Without microbes, we couldn't eat or breathe. Understanding microbes is vital to understanding the past and the future of ourselves and our planet. Archaea look and act a lot like bacteria. So much so that until the late 1970s, scientists assumed they were a kind of "weird" bacteria.

Then microbiologist Carl Woese devised an ingenious method of comparing genetic information showing that they could not rightly be called bacteria at all. Their genetic recipe is too different. So different Woese decided they deserved their own special branch on the great family tree of life, a branch he dubbed the Archaea.

Archaea comes from the Greek word meaning "ancient." An appropriate name, because many archaea thrive in conditions mimicking those found more than 3.5 billion years ago. Back then, the earth was still covered by oceans that regularly reached the boiling point — an extreme condition not unlike the hydrothermal vents and sulfuric waters where archaea are found today.

Some scientists consider archaea living fossils that may provide hints about what the earliest life forms on Earth were like, and how life evolved on our planet. In addition to superheated waters, archaea have been found in acid-laden streams around old mines, in frigid Antarctic ice and in the super-salty waters of the Dead Sea. A number of other extreme-living bacterial species also enjoy these conditions, too, such as the community of cyanobacteria and bacteria shown top right.

Foodborne illness or food poisoning is caused by consuming food contaminated with pathogenic bacteria, toxins, viruses, prions or parasites. Such contamination usually arises from improper handling, preparation or storage of food. Foodborne illness can also be caused by adding pesticides or medicines to food, or by accidentally consuming naturally poisonous substances like poisonous mushrooms or reef fish. Contact between food and pests, especially flies, rodents and cockroaches, is a further cause of contamination of food.

Some common diseases are occasionally foodborne mainly through the water vector, even though they are usually transmitted by other routes. These include infections caused by Shigella, Hepatitis A, and the parasites Giardia lamblia and Cryptosporidium parvum.

Scientists also use molecular tools to extract and compare bits of a particular kind of RNA from samples in order to determine if previously known or new microbes are present in a particular environment. This technique is widely used as a biomarker and for microbial ecology studies. It uses a particular kind of RNA known as 16S ribosomal RNA, or 16S rRNA.

Ribosomes are the gene-translating machines in all living things. When a gene on a piece of DNA is copied into a strand of messenger RNA and ferried out of the cell nucleus into the cell fluid, ribosomes there latch onto this mRNA. The ribosomes move along the mRNA strand, reading the code contained in its sequence of nucleotide bases (the As, Gs, Cs and Us, since U replaces T in RNA) and stringing the right amino acids together based on the code to build protein chains.

Pasteurization is typically associated with milk. There are two widely used methods to pasteurize milk: high temperature/short time (HTST), and ultra-high temperature (UHT). HTST is by far the most common method. Milk simply labelled "pasteurized" is usually treated with the HTST method, whereas milk labelled "ultra-pasteurized" must be treated with the UHT method. HTST involves holding the milk at a temperature of 161.5 degrees F (or 72 C) for at least 15 seconds. UHT involves holding the milk at a temperature of 280 degrees F (or 138 C) for at least two seconds.

Pasteurization methods are usually standardized and controlled by national food safety agencies. These agencies require milk to be HTST pasteurized in order to qualify for the "pasteurized" label. There are different standards for different dairy products, depending on the fat content and the intended usage. For example, the pasteurization standards for cream differ from the standards for fluid milk, and the standards for pasteurizing cheese are designed to preserve the phosphatase enzyme, which aids in curing the cheese.

The HTST pasteurization standard was designed to achieve a 5-log (approximately one million-fold) reduction in the number of viable microorganisms in milk. This is considered adequate for destroying almost all yeasts, mold, and common spoilage bacteria and also to ensure adequate destruction of common pathogenic heat-resistant organisms (including particularly Mycobacterium tuberculosis, which causes tuberculosis and Coxiella burnetii, which causes Q fever). HTST pasteurization processes must be designed so that the milk is heated evenly, and no part of the milk is subject to a shorter time or a lower temperature.

HTST pasteurized milk typically has a refrigerated shelf life of two to three weeks, whereas ultra pasteurized milk can last much longer when refrigerated, sometimes two to three months. When UHT pasteurization is combined with sterile handling and container technology, it can even be stored unrefrigerated for long periods of time.

IDENTIFICATION OF LACTIC ACID BACTERIA

Direct detection and identification of lactic acid bacteria in a food processing plant and in meat products using denaturing gradient gel electrophoresis. Takahashi H et al. established a novel system using denaturing gradient gel electrophoresis (DGGE) to quickly identify bacteria known to be responsible for spoilage in meat processing plants and meat products. We extracted bacterial DNA from swabbed samples at various locations in the plant and from meat products and performed PCR amplification, targeting 16S rDNA from the dominant organisms.

The amplification products were subjected to DGGE, and the contaminating bacteria in the meat products and the plant were analyzed. This analysis indicated that lactic acid bacteria and spoilage-causing bacteria are widely distributed within the meat processing plant. We developed molecular size markers to identify the dominant organisms obtained from the plant and meat products. The establishment of the present method allows quick and simple identification of bacteria causing the possible deterioration of products and contamination and thus permits constant monitoring of any harmful bacteria within meat processing plants.

Control of Listeria monocytogenes on frankfurters with antimicrobials in the formulation and by dipping in organic acid solutions; Barmpalia IM et al.; The antilisterial activity of sodium lactate (SL) and sodium diacetate (SD) was evaluated in a frankfurter formulation and in combination with a dipping treatment into solutions of lactic acid or acetic acid after processing and inoculation.

Pork frankfurters were formulated with 1.8% SL or 0.25% SD or combinations of 1.8% SL with 0.25 or 0.125% SD. After processing, frankfurters were inoculated (2 to 3 log CFU/cm^2) with a 10-strain composite of Listeria monocytogenes and left undipped or were dipped (2 min) in 2.5% solutions of lactic acid or acetic acid (23 +/- 2 degrees C) before vacuum packaging and storage at 10 degrees C for 40 days. Total microbial populations and L. monocytogenes, lactic acid bacteria, and yeasts and molds were enumerated during storage.

Sensory evaluations also were carried out on frankfurters treated and/or formulated with effective antimicrobials. The combination of 1.8% SL with 0.25% SD provided complete inhibition of L. monocytogenes growth throughout storage. Dipping in lactic acid or acetic acid reduced initial populations by 0.7 to 2.1 log CFU/cm^2, but during storage (12 to 20 days), populations on dipped samples without antimicrobials in the formulation reached 5.5 to 7.9 log CFU/cm^2. For samples containing single antimicrobials and dipped in lactic acid or acetic acid, L. monocytogenes growth was completely inhibited or reduced over 12 and 28 days, respectively, whereas final populations were lower ($P < 0.05$) than those in undipped samples of the same formulations.

Bactericidal effects during storage (reductions of 0.6 to 1.0 log CFU/ cm^2 over 28 to 40 days) were observed in frankfurters containing combinations of SL and SD that were dipped in organic acid solutions. Inclusion of antimicrobials in the formulation and/or dipping the product into organic acid solutions did not affect ($P > 0.05$) the flavor and overall acceptability of products compared with controls.

Acharya M.R et al., compared the effect of vacuum-condensed and ultrafiltered milk on cheddar cheese. The objective of this study was to compare the effects of vacuum-condensed (CM) and ultrafiltered (UF) milk on some compositional and functional properties of Cheddar cheese. Five treatments were designed to have 2 levels of concentration (4.5 and 6.0% protein) from vacuum-condensed milk (CM1 and CM2) and ultrafiltered milk (UF1 and UF2) along with a 3.2% protein control. The samples were analyzed for fat, protein, ash, calcium, and salt contents at 1 wk. Moisture content, soluble protein, meltability, sodium dodecyl sulfate-PAGE, and counts of lactic acid bacteria and nonstarter lactic acid bacteria were performed on samples at 1, 18, and 30 wk. At 1 wk, the moisture content ranged from 39.2 (control) to 36.5% (UF2).

Fat content ranged from 31.5 to 32.4% with no significant differences among treatments, and salt content ranged from 1.38 to 1.83% with significant differences. Calcium content was higher in UF cheeses than in CM cheeses followed by control, and it increased with protein content in cheese milk. Ultrafiltered milk produced cheese with higher protein content than CM milk. The soluble protein content of all cheeses increased during 30 wk of ripening. Condensed milk cheeses exhibited a higher level of proteolysis than UF cheeses. Sodium dodecyl sulfate-PAGE showed retarded proteolysis with increase in level of concentration.

The breakdown of alphas1- casein and alphas1-I-casein fractions was highest in the control and decreased with increase in protein content of cheese milk, with UF2 being the lowest. There was no significant degradation of beta-casein. Overall increase in proteolytic products was the highest in control, and it decreased with increase in protein content of cheese milk. No significant differences in the counts of lactic starters or nonstarter lactic acid bacteria were observed. Extent as well as method of concentration influenced the melting characteristics of the cheeses.

Melting was greatest in the control cheeses and least in cheese made from condensed milk and decreased with increasing level of milk protein concentration. Vacuum condensing and ultrafiltration resulted in Cheddar cheeses of distinctly different quality. Although both methods have their advantages and disadvantages, the selection of the right method would depend upon the objective of the manufacturer and intended use of the cheese.

BACTERIA

Bacteria can be found virtually everywhere. They are in the air, the soil, and water, and in and on plants and animals, including us. A single teaspoon of topsoil contains about a billion bacterial cells (and about 120,000 fungal cells and some 25,000 algal cells). The human mouth is home to more than 500 species of bacteria.

Some bacteria (along with archaea) thrive in the most forbidding, uninviting places on Earth, from nearly-boiling hot springs to super-chilled Antarctic lakes buried under sheets of ice. Microbes that dwell in these extreme habitats are aptly called extremophiles.

Like dinosaurs, bacteria left behind fossils. The big difference is that it takes a microscope to see them. And they are older. Bacteria and their microbial cousins the archaea were the earliest forms of life on Earth. And may have played a role in shaping our planet into one that could support the larger life forms we know today by developing photosynthesis.

Cyanobacteria fossils date back more than 3 billion years. These photosynthetic bacteria paved the way for today's algae and plants. Cyanobacteria grow in the water, where they produce much of the oxygen that we breathe. Once considered a form of algae, they are also known as blue-green algae.

The human body consists of millions of different cells. A bacterium consists of a single cell. A bacterium's genetic information is contained in a single DNA molecule suspended in a jelly-like substance called cytoplasm. In most cases, this and other cell parts are surrounded by a flexible membrane that is itself surrounded by a tough, rigid cell wall. A few species, such as the mycoplasmas, don't have cell walls.

Even though bacteria have only one cell each, they come in a wide range of shapes, sizes, and colors. Does a bacterium's cell wall, shape, way of moving, and environment really matter? Whether a bacteria has a thin or a thick cell wall determines what antibiotic will work against it. If you've ever been sick and waited for the results of a culture and sensitivity test, you may have heard the terms "Gram-positive" or "Gram-negative."

Bacteria with thick cell walls retain dye from a cell-staining method developed by Christian Gram; bacteria with thin walls do not. Knowing the difference can and does save lives, time and money, and ensures that you or your loved one is getting the best and most effective treatment.

Some bacteria look like little balls (Micrococcus) while others appear like tangled strings or corkscrews (Leptospira) under a microscope. Others look like medicine capsules (Salmonella) or segmented ribbons (Cyanobacteria) or sticks. Still others look like fat commas (Vibrios).

Some bacteria are stalked (Caulobacter) while others have buds (Rhodomicrobium). Some have sheaths

(Sphaerotilus) while others don't. Bacteria like mycoplasmas that lack a hard cell wall don't have any particular shape at all.

Just like in animals, where size ranges from the giant blue whale to the tiny gnat, bacteria vary from 1 millimeter in diameter at the largest end of the scale to 20 nanometers in length at the smallest. The largest bacteria found so far can actually be seen without the use of a microscope (Thiomargarita namibiensis and Epulopiscium fischelsoni). The smallest known bacteria are so tiny that they were once thought to be viruses (Mycoplasmas).

Some bacteria have hair- or whip-like appendages called flagella used to 'swim' around. Others produce thick coats of slime and 'glide' about. Some stick out thin, rigid spikes called fimbriae to help hold them to surfaces. Some contain little particles of minerals that orient with the planet's magnetic fields to help the bacteria figure out whether they're swimming up or down.

Fungi are eukaryotic organisms. This means that their DNA-containing chromosomes are enclosed within a nucleus inside their cells. (The chromosomes of bacteria and archaea are not walled off inside nuclei, making them prokaryotic organisms.)

Many decades ago, scientists thought that fungi were primitive kinds of plants. New studies looking at the DNA of fungi have confirmed that these organisms are not plants. Unlike plants, fungi do not make their own food energy via photosynthesis, but dine on organic matter like rotting leaves, wood, and other debris, or upon the tissues of living plants and animals.

Fungi, along with bacteria, are the planet's major composters and recyclers. In addition to the standard HTST and UHT pasteurization standards, there are other lesser-known pasteurization techniques. The first technique, called "batch pasteurization", involves heating large batches of milk to a lower temperature, typically 155 degrees fahrenheit (or 68 C). The other technique is called higher-heat/shorter time (HHST), and it lies somewhere between HTST and UHT in terms of time and temperature.

The batch pasteurization step, which is cheap at a large scale, is often performed prior to standard pasteurization. Batch pasteurized milk is often called "raw milk" or, confusingly, "unpasteurized milk". It cannot be called "pasteurized", even though a significant number of pathogens are destroyed during the process.

In recent years, there has been some consumer interest in raw milk products, due to perceived health benefits. Advocates of raw milk maintain, correctly, that vitamins and nutrients survive much better in milk that has not been pasteurized. They also maintain that organic raw milk (most retail raw milk is also organic) is less likely to contain harmful pathogens due to better husbandry in organic dairy herds. This may be true, but it has not been proven.

However, doctors (and even most raw milk advocates) acknowledge that certain people (e.g. pregnant or breast-feeding mothers, those undergoing immunosuppression treatment for cancer, organ transplant or autoimmune diseases, and those who are immunocompromised due to diseases like AIDS) should not risk consumption of raw milk.

In fact, some doctors suggest that babies and breast-feeding mothers avoid all but UHT pasteurized dairy products. In Africa, it is common to boil milk whenever it is harvested. This intense heating greatly changes the flavor of milk, which the people in Africa are accustomed to. Although fungi may seem like a nuisance when they dine in your fruit bowl or refrigerator, their ability to degrade some of the toughest organic materials, including tree wood and insect exoskeletons, means that our planet is not cluttered with a mass of debris. Fungi dine at home, eating whatever they're growing on. Fungi secrete digestive enzymes in order to break down complex food sources, such as animal corpses and tree stumps, into smaller components they can absorb.

Algae are plant-like microorganisms that preceded plants in developing photosynthesis, the ability to turn sunlight into energy. Algae cells contain light-absorbing chloroplasts and produce oxygen through photosynthesis. Although plants generally get the credit for producing the oxygen we breathe, some 75% or more of the oxygen in the planet's atmosphere is actually produced by photosynthetic algae and cyanobacteria.

Algae also play an important role as the foundation for the aquatic food chain. All higher aquatic life forms depend either directly or indirectly on microscopic gardens of algae. Most unicellular algae live in water, some dwell in moist soil, and others join with fungi to form lichens.

Green algae: The most clearly plant-like algae, this species gets its namesake hue from high levels of chlorophyll. Their cell walls are made up of cellulose, the same material that makes up the cell walls in larger, multicellular plants. Like plants, they store the food they make through photosynthesis as starches. Growing in large masses, these algae can form visible layers of slick, green scum on the surfaces and sides of ponds, puddles or damp soil.

Fossil records suggest that the first green algae originated 500 to 600 million years ago. Early algae probably gave rise to multicellular plants. Dinoflagellates have long whip-like structures called flagella that let them turn, maneuver and spin about through the water. About 90% of these algae dwell in the ocean.

Some species glow in the dark in a process called bioluminescence. These species contain a compound called luciferin (the same compound found in fireflies). The glow increases markedly if the algae cells are agitated, as when a ship churns through the water.

About half the species of dinoflagellates are photosynthetic; the other half are predators that attack bacteria, algae, and even fish.

Dinoflagellate neurotoxins can concentrate in the bodies of shellfish and fish that eat the algal cells, in turn causing people who eat these seafoods to come down with illnesses such as paralytic shellfish poisoning and ciguatera (a combination of gastrointestinal, neurological, and cardiovascular disorders.)

So-called "red tides" occur when enormous blooms of trillions of dinoflagellates are triggered by an upwelling of nutrients from the water's depths during warmer seasons. The population of dinoflagellates can jump to more than 20 million cells per liter of sea water along some coasts during these blooms, turning the water a reddish hue.

The name protozoa means "first animals." As the principal hunters and grazers of the microbial world, protozoa play a key role in maintaining the balance of bacterial, algal, and other microbial life. They also are themselves an important food source for larger creatures and the basis of many food chains.

Protozoa have been found in almost every kind of soil environment from peat bogs to arid desert sands. They teem in the deep sea as well as near the surface of waters, and can be found even in frigid Arctic and Antarctic waters.

Some species of protozoa are part of the normal microbial flora of animals, and live in the guts of insects and mammals, helping to break down complex food particles into simpler molecules. A very small number of species cause disease in people, including Plasmodium vivax, which causes malaria.

Bacterial infection is the most common cause of food poisoning. Symptoms for bacterial infections are delayed because the bacteria need time to multiply. They are usually not seen until 12-36 hours after eating contaminated food.

Common bacterial foodborne pathogens are:

Aeromonas hydrophila, Aeromonas caviae, Aeromonas sobria Bacillus cereus Brucella spp. Campylobacter jejuni which causes Guillain-Barré syndrome Corynebacterium ulcerans Coxiella burnetii or Q fever Crohn's disease Escherichia coli O157:H7 enterohemorrhagic (EHEC) which causes hemolytic-uremic syndrome Escherichia coli - enteroinvasive (EIEC) Escherichia coli - enteropathogenic (EPEC) Escherichia coli - enterotoxigenic (ETEC) Listeria monocytogenes Plesiomonas shigelloides Salmonella spp. Shigella spp. Streptococcus Vibrio cholerae, including O1 and non-O1 Vibrio parahaemolyticus Vibrio vulnificus Yersinia enterocolitica and Yersinia pseudotuberculosis

The four main subgroups of protozoa are the ciliates, the flagellates, the sarcodina, and the apicomplexans.

A few ciliates can grow up to 2 millimeters in length, big enough to be seen without a microscope.

Flagellates Similarly complex single-celled organisms, flagellates have whip-like appendages called flagella sticking out of their cells.

The flagella are used for locomotion and to direct food particles or cells into the organism's mouth-like opening. Flagellates dine on bacteria, algae, and other protozoa.

Several well-known flagellates cause parasitic diseases, such as trypanosomes that cause sleeping sickness, and Giardia lamblia, a parasite found in mountain streams and rivers that causes severe gastrointestinal distress.

Heliozoa, radiozoa, and forams tend to be passive grazers and predators, relying on suitable food swimming or drifting past to come into contact with their pseudopods.

Several species in the sarcodina group, including some species of amoebas, cover themselves with protective shell-like coverings called tests. These tests are stippled with many small and large openings through which water can flow in and out and through which the pseudopods protrude.

The tests of radiozoa are made up of silica (the same substance in diatom cell walls) and can form very intricate, lacy designs that may be studded with long spines that increase buoyancy and ward off predators.

The tests of forams are made up of sand grains or organic compounds. These can become quite large, the biggest reaching a little over 2 inches in diameter, making them some of the largest single-celled organisms known. When forams die, their tests sink and accumulate in large batches; the Great Pyramids of Egypt are built from sandstone composed largely of fossilized giant Nummulites, an ancient kind of foram. The famous White Cliffs of Dover are limestone cliffs formed from the skeletal remains of forams.

Apicomplexans These protozoa are obligatory intracellular parasites: they must spend at least part if not all of their life cycle in a host animal.

Apicomplexans are characterized by the presence of special organelles (tiny organ-like structures) located at the tips (apices) of the cells. These organelles contain enzymes that punch through, slice open and otherwise penetrate host tissues.

The best known apicomplexan is Plasmodium, the agent that causes malaria. Plasmodium spends part of its life cycle in mosquitoes and the other part in human hosts where it ultimately infects and ruptures blood cells in large numbers.

Another familiar apicomplexan is Cryptosporidium parvum. This water-borne parasite forms extremely durable cyst-like structures that enable it to survive UV radiation and sometimes chlorine in swimming pools and treated water. An outbreak of Cryptosporidium in

Milwaukee's drinking water supply in 1993 killed 50 people and sickened more than 400,000.

Probably the best known and most deadly case of contamination in the U.S. in recent years happened in Milwaukee in 1993, when the municipal water supply was contaminated by Cryptosporidium, an intestinal protozoan. At least 50 people died, and some 400,000 people became ill, 4,000 badly enough to be hospitalized.

In 1973, a Dallas resident went out to the backyard only to stumble upon a reddish, jelly-like mass pulsating in the grass. News reports on the discovery claimed that a "new life form" had been found, and many people couldn't help recalling the cult classic sci-fi thriller The Blob.

Scientists called to the scene, however, put any fears of menacing goo or alien creatures to rest by identifying the mass as an unusually large (46 centimeters or more than 14 inches in diameter) plasmodial slime mold.

Slime molds were once considered fungi, but unlike fungi, they can move, and their cell membranes are made of different stuff. Slime molds are made up of individual cells that form an aggregate mass. In their visible, aggregate states, they look like blobs, gooey or foamy masses, spilled jelly, or even dog vomit. They may be bright orange, red, yellow, brown, black, blue, or white.

These large masses act like giant amoebas, creeping slowly along and engulfing food particles along the way. If a slime mold aggregate is diced up, the pieces will pull themselves back together. The blobs can navigate and avoid obstacles and if a food source is placed nearby, they seem to sense it and head unerringly for it. j, a. There are two kinds of slime molds. Plasmodial slime molds (the most common kind) share one big cell wall that surrounds thousands or millions of nuclei. Proteins called microfilaments act like tiny muscles that enable the mass to crawl at rates of about 1/25th of an inch per hour.

As long as there is enough food and moisture, the mass thrives. But when food and water are scarce, the mass separates into smaller blobs. The Plasmodium forms stalks topped by sphere-like fruiting bodies that contain spores that are carried by the rain or wind to new locations.

Cellular slime molds also produce spores, but these germinate into amoeba-like cells. The cells happily go their individual ways, as long as food and water are available. When nutrients and moisture are scarce, individual cells send out a chemical beacon to attract other cells of the same species. The cells join up to form a mass that looks and acts like a slug to take them to a more favorable location.

Cells in cellular slime molds retain their individual cell walls when they form a mass, so the visible slug is actually a collection of hundreds of thousands of individual cells joined together.

Slime molds eat decaying vegetation, bacteria, fungi, and even other slime molds. They are most commonly found in forests.

Milk pasteurization standards have been subject to increasing scrutiny in recent years, due to the discovery of pathogens that are both widespread and heat resistant (able to survive pasteurization in significant numbers). Researchers have developed more sensitive diagnostics, such as real-time PCR and improved culture methods, that have enabled them to identify pathogens in pasteurized milk.

COMPONENTS OF MICROBIOLOGICAL CRITERION

The components of a microbiological criterion are described and some of the potential contributions of criteria to food safety and quality are discussed. That microbiological criteria should be established and implemented only when there is a need and when the criterion can be shown to be effective and practical. There are a number of factors that will determine the foods for which criteria related to safety might be applied. Of special importance is epidemiological evidence that the food in question is a significant vehicle of disease. Also important are various hazard considerations which include (1) susceptibility of the food to contamination by pathogens, (2) the opportunity for the survival of pathogens, (3) the likelihood of microbial growth during manufacture, storage, distribution, and preparation for serving, (4) whether or not the food is to be cooked prior to serving, and (5) the susceptibility of probable consumers to infectious agents or toxins.

Microorganisms suitable as components of microbiological criteria (standards, guidelines, specifications) have been placed in two categories-pathogens and indicator organisms. Pathogens suitable for this purpose are those likely to be found in a food or ingredient that thereby becomes a potential vehicle for transmission of the organism or its toxin to consumers. Indicator organisms are those whose presence in a food indicates that a pathogen or its toxin of concern may be present, that faulty practices occurred that may adversely affect safety or shelf-life of the product, or that the food or ingredient is unsuited for an intended use.

Although the list of foodborne diseases and the microbial agents that cause them is long, only about 20 are known to be transmitted by foods with a consequence and/or frequency serious enough to cause concern. Indicator organisms and agents were evaluated relative to their use for the assessment of numbers of microorganisms and/or microbial activity, potential human or fecal contamination or potential presence of pathogens, and post-heat processing contamination. Each of these organisms, in addition to agents, was discussed relative to its importance, its status and limitation of method of detection or enumeration, and its suitability as a part of a microbiological criterion.

It is recognized that the sampling plan and decision criteria are essential components of a microbiological criterion. They should be based on sound statistical concepts. The International Commission on Microbiological Specifications for Foods (ICMSF) concept of relating the stringency of the sampling plan to the degree of hazard of the food was endorsed by the subcommittee as an effective means in the selection process. Whenever applicable, attributes sampling plans such as the ICMSF 2- and 3-class plans are recommended for microbiological criteria of foods and food ingredients in the United States.

The action taken when the limit of a criterion is exceeded relates to the purpose for which the criterion was established. Actions that may be taken under the following circumstances: (1) evidence of existence of a direct health hazard, (2) evidence that a direct health hazard could develop, (3) indications that a product was not produced under conditions assuring safety, (4) indications that a raw material may adversely affect shelf-life, and (5) evidence that a critical control point is not under control.

MICROBIOLOGICAL CRITERIA IN FOOD INGREDIENTS

Dairy Products

Microbiological criteria play an important role as part of overall quality assurance programmes applied by both industry and regulatory authorities. Microbiological safety of dairy products can be assured only through (1) pasteurization or more severe heat treatments, (2) prevention of post-heat treatment contamination and, for certain products, (3) end-product testing for appropriate pathogens or toxins. Microbiological criteria for most dairy products are included as an integral part of documents that present in detail the FDA and USDA dairy products control programmes.

Cheese is the other product for which greater attention to finished-product testing for a pathogen or its toxin is indicated. Of primary concern is S. aureus, especially its enterotoxins. The hazard is limited primarily to hard varieties, e.g., Cheddar and similar types and Swiss or Emmenthaler. This hazard has been discussed and suggestions have been made for more rigorous application of microbiological criteria in programmes to control this hazard. The continuous problem of the hazard of consuming raw milk and fresh cheese made from raw milk have been given particular attention.

Outbreaks of milkborne disease caused by drinking raw milk purchased legally continue to occur in the United States with regularity. Brucellosis due to consumption of fresh cheese made from raw milk and enterocolitis caused by certain pathogenic strains of E. coli are of recent occurrence. A strong position has been taken pointing out that health professionals have a responsibility to encourage efforts to see that the public and policymakers are kept adequately informed so that public policy on raw milk and dairy products made from raw milk is compatible with scientific knowledge and protection of the public's health.

Raw Meats and Poultry

The causes and contributing factors of foodborne illness from meats and poultry have been thoroughly examined. The subcommittee could not establish a need for microbiological standards for pathogens in raw meat and poultry. Efforts to reduce foodborne illness caused by cooked meat and poultry should be directed to improved education, e.g., recognition of the potential presence of pathogens in raw animal foods and proper handling of raw and processed animal foods. In view of this need, renewed efforts to educate all persons that handle raw and processed animal foods of potential risks and proper food-handling practices are of the highest priority.

Factors relating to achieving optimum shelf-life of raw meat and poultry are discussed. These include processing and storage conditions. The application of microbiological guidelines to monitor these operations is indicated. Present inspection practices of food-processing facilities by regulatory agencies need a more cost-effective approach by application of the HACCP system. Because the vast majority of foodborne disease is the direct result of mishandling of foods in food service operations and homes, the extent of regulatory inspection of food service operations is disproportionate compared with that of food-processing facilities. Therefore, emphasis of inspection should be directed more to those areas where mishandling of animal foods occurs more frequently, namely food service operations.

Fish, Molluscs, and Crustaceans

Although strict adherence to the recommendations of the National Shellfish Sanitation Programme (NSSP) generally has resulted in the production of safe shellfish, there is need for more effective measures for monitoring the safety of these products, particularly with reference to the presence of viruses that may cause illness in humans.

Intoxication (ciguatera and scombroid and paralytic shellfish poisoning [PSP] resulting from consumption of fish and shellfish is a major concern. The regulatory limits for histamine in tuna and PSP toxin in raw shellfish are recognized as useful in the control of these foodborne diseases. Relative to ciguatera there is a need for a simple reliable test to detect toxic fish. The quality of raw fish can best be assured through application of the HACCP approach by sensory inspection of raw materials, proper temperature control, sanitation of equipment, and proper handling of product by employees.

Thus, this packaging-storage method is not recommended for raw fish at the present time -there is need for research to examine the safety of this practice. Except for the NSSP criteria for shellfish, many of the microbiological criteria for fish and seafoods at the state or local level are not based on sound data or experience and are impractical from the standpoint of compliance or enforcement.

Processed Animal Products

Though such products as ground meat, poultry parts, and fish fillets are classified as processed, the safety and spoilage problems involved are the same as discussed above relative to raw animal products.

In the present context, processed animal products are foods that are produced through the application of various physical and/or chemical treatments to raw animal products. Such treatments as pasteurization, curing, salting, and acidulation (including fermentation) tend to destroy or otherwise control the development of the microorganisms that normally spoil the raw products. Likewise, to a greater or lesser extent, these treatments destroy or prevent the growth of foodborne disease agents.

These products vary considerably in their stability. Country cured hams and dry sausages may be stored for long periods at ambient temperatures without spoilage, whereas luncheon meats, cooked poultry, and smoked fish are highly perishable. Similarly, some of these products, if contaminated and subject to temperature abuse, readily support the growth of toxigenic organisms. Ham, for example, has historically been the most important cause of staphylococcal foodborne illness. Others such as dried meats, poultry, and fish will not support the growth of pathogens but may create hazards if abused after reconstitution with water.

The application of the HACCP system to the control of the safety and quality of these products is discussed. In many cases microbiological criteria play a role and where they apply they have been identified; but more often, physical and chemical measurements are the most effective means of monitoring critical control points. Three products in this category have in recent years caused multiple outbreaks of foodborne illness, namely smoked fish (botulism), dry sausage (staphylococcal intoxication), and cooked roast beef (salmonellosis).

Eggs and Egg Products

Microbiological criteria play no role in the control of shell egg quality. Defective shell eggs are essentially eliminated from commerce by sorting and candling. Microbiological guidelines are useful in the monitoring of critical control points in the production of egg products, e.g., equipment sanitation and environmental monitoring. The pasteurizing step is more effectively monitored by observing time/temperature relationships.

Fruits and Vegetables

Raw fruits and vegetables as marketed may harbor large populations of viable microorganisms, but they have not presented a serious public health problem in the United States and consequently the application of microbiological criteria to these foods is not recommended. In the processing of fruits and vegetables for canning, drying, and freezing, there are opportunities for microbial buildup on equipment and growth in the food. Guidelines that limit counts of microorganisms or utilize microscopic counts for molds are recommended as a means of monitoring critical control points and promoting good processing conditions.

Fruit Beverages

Because of their high acidity, fruit juice beverages have rarely been vehicles for foodborne pathogens. Guidelines that limit numbers of viable organisms or levels of metabolic products can be useful to monitor critical control points in the processing line. Specifications that restrict spore counts of heat-resistant molds are recommended as a means to minimize spoilage outbreaks of pasteurized fruit beverages.

Canned Foods

The production of low-acid, acid, and water activity-controlled canned foods is regulated by the FDA. Regulations governing this programme are contained in the Code of Federal Regulations. These regulations embody the HACCP approach to the control of safety and quality. Canned products were the first foods to which the application of HACCP was mandated by regulation. The success of this programme should serve as a model to other segments of the food industry.

Though finished-product microbiological criteria play no role in assuring the acceptability of finished products, guidelines are used in monitoring critical control points including equipment cleaning and sanitizing, microbial buildup during operations, and the microbiological quality of chill water. Further, specifications are applied to critical raw materials to assure that the level of heatresistant sporeformers does not exceed acceptable limits.

Cereals and Cereal Products

The usefulness of criteria for grains, pasta products, and pastries was considered, as were public health problems associated with these foods. The presence of mycotoxins on grains, contamination of soy and pasta products with salmonellae, and the growth of Staphylococcus aureus in certain cream-filled pastries were identified as the major problems. The use of microbiological criteria in support of HACCP programmes is recommended for these products.

Fats and Oils

Fats and oils in the presence of moisture and other essential nutrients are subject to degradation by a variety of microorganisms. Such a condition is provided by several processed foods in which fats and oils are major ingredients. Mayonnaise, salad dressings, peanut butter, margarine, and butter are of concern. For each of these products, the application of each or all of the three types of microbiological criteria is useful and in some instances essential to safety.

For example, with respect to peanut butter routine testing of raw peanuts for aflatoxin and of the finished product for Salmonella is essential. The characteristics of these products that relate to either microbial growth or inhibition of growth in them are discussed, as are significant sources of contamination with either spoilage organisms or pathogens. Applications of microbiological guidelines for monitoring critical control points of manufacture are described. The need for routine testing of peanut butter for Salmonella has been emphasized. The subcommittee recommends amending the standards of identity for mayonnaise and salad dressings to include a specific requirement for a pH of 4.1 or below in the aqueous phase of these products.

Sugar, Cocoa, Chocolate, and Confectioneries

The major microbiological concerns for these foods are the presence of bacterial spores and yeasts in sugar and of salmonellae in cocoa and chocolate. Specifications that limit thermophilic spores in sugar for canning; yeasts, molds, and mesophilic bacteria in sugar for bottled beverages; and osmophilic yeasts for confectionery products are appropriate. The ingredients of confectionery and similar ready-to-eat foods should be free of enteric pathogens, in particular salmonellae.

Spices

Microbiological concerns are governed by end use. Spices to be used as table condiments should be free of salmonellae. Spices containing high numbers of sporeforming bacteria may cause spoilage during the heat processing of certain luncheon meats; those containing thermophilic sporeformers may cause spoilage of canned foods. In these instances, spices as ingredients constitute a critical control point and are generally monitored via specifications. Ethylene oxide and irradiation treatments are appropriate control procedures when necessary.

Yeasts

Salmonellae in yeasts, especially in nutritional yeast that will be eaten without further heating, are a major concern. HACCP programmes in the yeast industry must include routine testing for this pathogen. Microbiological guidelines are useful in the yeast industry for monitoring good manufacturing practices. Microbiological testing is also useful for the detection of contaminants that may cause spoilage of products containing yeasts, e.g., rope spores.

Formulated Foods

This broad category of products includes commercially prepared ready-tocook or ready-to-eat foods containing ingredients from two or more food commodity categories. Beef and chicken potpies, the various packaged meals or "dinners," meat and seafood salads, pizzas, dried infant formulae, dry soup mixes, and cake mixes are only a few examples of the host of such products that are available to the public. It has been emphasized that no useful single microbiological criterion can be developed. Rather, the potential hazards of each product or group of products must be identified.

Appropriate programmes that embody the HACCP system including microbiological criteria (particularly guidelines as applied at critical control points in manufacture) must be individually designed if microbial growth and resulting hazard to health or of spoilage are to be prevented. Attention has been drawn to regulations of several states that specify a single criterion for all products within diverse groups such as salads. The deficiency of such criteria is presented.

Nuts

With the exception of coconut, the water activity of most nuts is sufficiently low to preclude bacterial growth. Because the growth of molds with the production of mycotoxins is a problem with peanuts and certain tree nuts, testing for these toxins should be part of a HACCP programme. Nonsterile coconut should be free of salmonellae. A standard for E. coli on tree nuts has been imposed by the FDA. The presence of this organism is interpreted as an indication of the presence of filth. The subcommittee is not aware of epidemiological evidence that suggests a relationship between the presence of E. coli and enteric pathogens in nuts.

FERMENTATION

Fermentation, initiated by the action of micro-organisms occurs naturally and is often part of the process of decay, especially in fruits and vegetables. However, fermentation can be controlled to give beneficial results. Fermentation is a relatively efficient, low energy preservation process, which increases the shelf life and decreases the need for refrigeration or other form of food preservation technology. It is therefore a highly appropriate technique for use in developing countries and remote areas where access to sophisticated equipment is limited.

Fruits

In terms of food processing, fruits are nearly all acidic and are therefore called 'high acid' foods. The acidity naturally controls the type of organism that can grow in fruits, with yeasts and moulds being the only spoilage organisms likely to be found on fruit products. The acidity level of tropical fruits, such as banana, mango and papaya, decreases as the fruit ripens. With respect to food processing and preservation, it is probably this definition of a fruit that will be most useful.

Vegetables

In terms of food processing, vegetables are classified as 'low acid' foods due to their lower levels of acidity. Low acid foods are more prone to deterioration by micro-organisms and can in fact provide an ideal substrate for food poisoning organisms when in a moist environment. Low acid foods can be safely preserved by making them more acidic, either through pickling or salting or drying.

Agro-processing

Agro-processing describes the transformation of agricultural produce into a different physical or chemical state. The term *agro* encompasses a wide range of food and non-food agricultural products. The term *food processing* applies only to products which are suitable for human consumption. *Agro processing* applies to any of the numerous activities that take place in the chain of events between harvest or slaughter of the raw material and production of the final product. It covers a range of processes with varying degrees of complexity and technical input to suit the individual situation. Different treatments range from the relatively simple processes of rice husking, drying and grinding to the more complex transformation of oilseeds into margarine and essential oil distillation.

Salometer

Salt present in a brine is expressed as degrees "Salometer" which is a percent saturation of sodium chloride by weight. A saturated solution of pure sodium chloride (100 degrees Salometer) contains 26.359g per 100ml at 15.5 degrees C. Therefore a 10 degree Salometer contains 2.64% salt by weight.

Biotechnology

Biotechnology can be described as the application of scientific and engineering principles to the processing of materials, for the provision of goods and services, through the use of biological systems and agents.

Complex Carbohydrate

A complex carbohydrate is one that is composed of long branched chains of single sugar units (glucose, fructose or galactose). Examples of complex carbohydrates include starch and cellulose.

Reducing Sugar

A reducing sugar is a sugar, which has reactive aldehyde or ketone groups. All simple sugars are reducing sugars. Sucrose, a common sugar, is not a reducing sugar.

Micro-organism/Microbe

Microbe and micro-organism are generic terms for the group of living organisms which are microscopic in size. Included in the definition are bacteria, viruses, moulds, yeasts and fungi.

Enzyme

An enzyme is a biological catalyst, which is used to facilitate and speed up reactions. It is a protein and requires a specific substrate to work on. Its working conditions are set within narrow limits – for example, optimum temperature, pH conditions and oxygen concentration. At temperatures above 42° C, mammalian proteins (and therefore enzymes) are denatured. However, certain bacterial enzymes are tolerant of a more diverse temperature range.

Hydrolysis

Hydrolysis is the splitting or breaking down of complex molecules by the action of enzymes or acid. For example the hydrolysis of starch and cellulose both yield simple glucose units.

Aerobic

With reference to micro-organisms, one which requires oxygen for survival.

Anaerobic

An anaerobic organism is one which does not require oxygen for survival.

Facultative anaerobe

A micro-organism which can adapt to exist with or without oxygen.

Microaerophilic

An organism which requires small amounts of oxygen for survival.

Water activity (a_w)

Water exists in two states within a cell – free and bound. Water activity is a measure of the amount of free water available for a potential reaction – microbial or enzymic.

Water activity is a measure of the free moisture in a product and is the quotient of the water vapour pressure of the substance divided by the vapour pressure of pure water at the same temperature. It is measured on a scale of 0 to 1.0 where 1.0 is the activity of pure water.

pH

pH is a measure of the hydrogen ion concentration. It is measured on a scale of 1 to 14, where 1 represents a high concentration of hydrogen ions (acidic) and 14 represents a low concentration (alkaline). The optimum pH for most micro-organisms is near neutral (pH 7.0). However, certain species are acid tolerant . Foods with a pH of 4.6 or lower are termed high acid foods. If the pH is above 4.6, it is a low acid food and is more prone to bacterial and fungal spoilage.

Microorganisms in Food Fermentation

The most common groups of micro-organisms involved in food fermentations are:

— Bacteria

— Yeasts

— Moulds

Bacteria

Several bacterial families are present in foods, the majority of which are concerned with food spoilage. As a result, the important role of bacteria in the fermentation of foods is often overlooked. The most important bacteria in desirable food fermentations are the *lactobacillaceae* which have the ability to produce lactic acid from carbohydrates. Other important bacteria, especially in the fermentation of fruits and vegetables, are the acetic acid producing *acetobacter* species.

Yeasts

Yeasts and yeast-like fungi are widely distributed in nature. They are present in orchards and vineyards, in the air, the soil and in the intestinal tract of animals. Like bacteria and moulds, yeasts can have beneficial and non-beneficial effects in foods. The most beneficial yeasts in terms of desirable food fermentation are from the *Saccharomyces* family, especially *S. cerevisiae*. Yeasts are unicellular organisms that reproduce asexually by budding. In general, yeasts are larger than most bacteria. Yeasts play an important role in the food industry as they produce enzymes that favour desirable chemical reactions such as the leavening of bread and the production of alcohol and invert sugar.

Moulds

Moulds are also important organisms in the food industry, both as spoilers and preservers of foods. Certain moulds produce undesirable toxins and contribute to the spoilage of foods. The *Aspergillus* species are often responsible for undesirable changes in foods. These moulds are frequently found in foods and can tolerate high concentrations of salt and sugar. However, others impart characteristic flavours to foods and others produce enzymes, such as amylase for bread making.

Moulds from the genus *Penicillium* are associated with the ripening and flavour of cheeses. Moulds are aerobic and therefore require oxygen for growth. They also have the greatest array of enzymes, and can colonise and grow on most types of food. Moulds do not play a significant role in the desirable fermentation of fruit and vegetable products.

When micro-organisms metabolise and grow they release by-products. In food fermentations the by-products play a beneficial role in preserving and changing the texture and flavour of the food substrate. For example, acetic acid is the by-product of the fermentations of some fruits. This acid not only affects the flavour of the final product, but more importantly has a preservative effect on the food. For food fermentations, micro-organisms are often classified according to these by-products.

The fermentation of milk to yoghurt involves a specific group of bacteria called the lactic acid bacteria (*Lactobacillus* species). This is a general name attributed to those bacteria which produce lactic acid as they grow. Acidic foods are less susceptible to spoilage than neutral or alkaline foods and hence the acid helps to preserve the product. Fermentations also result in a change in texture. In the case of milk, the acid causes the precipitation of milk protein to a solid curd.

Enzymic Activities in Fermentation

The changes that occur during fermentation of foods are the result of enzymic activity. Enzymes are complex proteins produced by living cells to carry out specific biochemical reactions. They are known as catalysts since their role is to initiate and control reactions, rather than being used in a reaction. Because they are proteinaceous in nature, they are sensitive to fluctuations in temperature, pH, moisture content, ionic strength and concentrations of substrate and inhibitors.

Each enzyme has requirements at which it will operate most efficiently. Extremes of temperature and pH will denature the protein and destroy enzyme activity. Because they are so sensitive, enzymic reactions can easily be controlled by slight adjustments to temperature, pH or other reaction conditions. In the food industry, enzymes have several roles - the liquefaction and saccharification of starch, the conversion of sugars and the modification of proteins. Microbial enzymes play a role in the fermentation of fruits and vegetables.

Nearly all food fermentations are the result of more than one micro-organism, either working together or in a sequence. For example, vinegar production is a joint effort between yeast and acetic acid forming bacteria. The yeast convert sugars to alcohol, which is the substrate required by the *acetobacter* to produce acetic acid.

Bacteria from different species and the various micro-organisms - yeast and moulds -all have their own preferences for growing conditions, which are set within narrow limits. There are very few pure culture fermentations. An organism that initiates fermentation will grow there until it's by-products inhibit further growth and activity. During this initial growth period, other organisms develop which are ready to take over when the conditions become intolerable for the former ones.

In general, growth will be initiated by bacteria, followed by yeasts and then moulds. There are definite reasons for this type of sequence. The smaller micro-organisms are the ones that multiply and take up nutrients from the surrounding area most rapidly. Bacteria are the smallest of micro-organisms, followed by yeasts and moulds. The smaller bacteria, such as *Leuconostoc* and *Streptococcus* grow and ferment more rapidly than their close relations and are therefore often the first species to colonise a substrate.

Desirable Fermentation

It is essential with any fermentation to ensure that only the desired bacteria, yeasts or moulds start to multiply and grow on the substrate. This has the effect of suppressing other micro-organisms which may be either pathogenic and cause food poisoning or will generally spoil the fermentation process, resulting in an end-product which is neither expected or desired. An everyday example used to illustrate this point is the differences in spoilage between pasteurised and unpasteurised milk. Unpasteurised milk will spoil naturally to produce a sour tasting product which can be used in baking to improve the texture of certain breads.

Pasteurised milk, however, spoils (non-desirable fermentation) to produce an unpleasant product which has to be disposed of. The reason for this difference is that pasteurisation (despite being a very important process to destroy pathogenic micro-organisms) changes the micro-organism environment and if pasteurised milk is kept unrefrigerated for some time, undesirable micro-organisms start to grow and multiply before the desirable ones.

In the case of unpasteurised milk, the non-pathogenic lactic acid bacteria start to grow and multiply at a greater rate that any pathogenic bacteria. Not only do the larger numbers of lactic acid bacteria compete more successfully for the available nutrients, but as they grow they produce lactic acid which increases the acidity of the substrate and further suppresses the bacteria which cannot tolerate an acid environment.

Most food spoilage organisms cannot survive in either alcoholic or acidic environments. Therefore, the production of both these end products can prevent a food from spoilage and extend the shelf life. Alcoholic and acidic fermentations generally offer cost effective methods of preserving food for people in developing countries, where more sophisticated means of preservation are unaffordable and therefore not an option.

The principles of microbial action are identical both in the use of micro-organisms in food preservation, such as through desirable fermentations, and also as agents of destruction via food spoilage. The type of organisms present and the environmental conditions will determine the nature of the reaction and the ultimate products. By manipulating the external reaction conditions, microbial reactions can be controlled to produce desirable results. There are several means of altering the reaction environment to encourage the growth of desirable organisms. These are discussed below.

Microbial Growth and Activity

There are six major factors that influence the growth and activity of micro-organisms in foods. These are moisture, oxygen concentration, temperature, nutrients, pH and inhibitors. The food supply available to the micro-organisms depends on the composition of the food on which they grow. All micro-organisms differ in their ability to metabolise proteins, carbohydrates and fats. Obviously, by manipulating any of these six factors, the activity of micro-organisms within foods can be controlled.

Moisture

Water is essential for the growth and metabolism of all cells. If it is reduced or removed, cellular activity is decreased. For example, the removal of water from cells by drying or the change in state of water (from liquid to solid) affected by freezing, reduces the availability of water to cells (including microbial cells) for metabolic activity. The form in which water exists within the food is important as far as microbial activity is concerned. There are two types of water - free and bound. Bound water is present within the tissue and is vital to all the physiological processes within the cell.

Free water exists in and around the tissues and can be removed from cells without seriously interfering with the vital processes. Free water is essential for the survival and activity of micro-organisms. Therefore, by removing free water, the level of microbial activity can be controlled. The amount of water available for micro-organisms is referred to as the water activity (a_w). Pure water has a water activity of 1.0. Bacteria require more water than yeasts, which require more water than moulds to carry out their metabolic activities.

Almost all microbial activity is inhibited below aw of 0.6. Most fungi are inhibited below aw of 0.7, most yeasts are inhibited below aw of 0.8 and most bacteria below aw 0.9. Naturally, there are exceptions to these guidelines and several species of micro-organism can exist outside the stated range. The water activity of foods can be changed by altering the amount of free water available. There are several ways to achieve this – drying to remove water; freezing to change the state of water from liquid to solid; increasing or decreasing the concentration of solutes by adding salt or sugar or other hydrophylic compounds (salt and sugar are the two common additives used for food preservation).

Addition of salt or sugar to a food will bind free water and so decrease the aw. Alternatively, decreasing the concentration will increase the amount of free water and in turn the aw. Manipulation of the aw in this manner can be used to encourage the growth of favourable micro-organisms and discourage the growth of spoilage ones.

Oxidation-Reduction Potential

Oxygen is essential to carry out metabolic activities that support all forms of life. Free atmospheric oxygen is utilised by some groups of micro-organisms, while others are able to metabolise the oxygen which is bound to other compounds such as carbohydrates. This bound oxygen is in a reduced form.

Micro-organisms can be broadly classified into two groups - aerobic and anaerobic. Aerobes grow in the presence of atmospheric oxygen while anaerobes grow in the absence of atmospheric oxygen. In the middle of these two extremes are the facultative anaerobes which can adapt to the prevailing conditions and grow in either the absence or presence of atmospheric oxygen.

Microaerophilic organisms grow in the presence of reduced amounts of atmospheric oxygen. Thus, controlling the availability of free oxygen is one means of controlling microbial activity within a food. In aerobic fermentations, the amount of oxygen present is one of the limiting factors. It determines the type and amount of biological product obtained, the amount of substrate consumed and the energy released from the reaction. Moulds do not grow well in anaerobic conditions, therefore they are not important in terms of food spoilage or beneficial fermentation, in conditions of low oxygen availability.

Temperature

Temperature affects the growth and activity of all living cells. At high temperatures, organisms are destroyed, while at low temperatures, their rate of activity is decreased or suspended. Micro-organisms can be classified into three distinct categories according to their temperature preference.

Nutritional Requirements

The majority of organisms are dependent on nutrients for both energy and growth. Organisms vary in their specificity towards different substrates and usually only colonise foods which contain the substrates they require. Sources of energy vary from simple sugars to complex carbohydrates and proteins. The energy requirements of micro-organisms are very high. Limiting the amount of substrate available can check their growth.

Hydrogen ion concentration (pH)

The pH of a substrate is a measure of the hydrogen ion concentration. A food with a pH of 4.6 or less is termed a high acid or acid food and will not permit the growth of bacterial spores. Foods with a pH above 4.6. are termed low acid and will not inhibit the growth of bacterial spores. By acidifying foods and achieving a final pH of less than 4.6, most foods are resistant to bacterial spoilage.

The optimum pH for most micro-organisms is near the neutral point (pH 7.0). Certain bacteria are acid tolerant and will survive at reduced pH levels. Notable acid-tolerant bacteria include *the Lactobacillus* and *Streptococcus* species, which play a role in the fermentation of dairy and vegetable products. Moulds and yeasts are usually acid tolerant and are therefore associated with spoilage of acidic foods.

Micro-organisms vary in their optimal pH requirements for growth. Most bacteria favour conditions with a near neutral pH (7). Yeasts can grow in a pH range of 4 to 4.5 and moulds can grow from pH 2 to 8.5, but favour an acid pH. The varied pH requirements of different groups of micro-organisms is used to good effect in fermented foods where successions of micro-organisms take over from each other as the pH of the environment changes. For instance, some groups of micro-organisms ferment sugars so that the pH becomes too low for the survival of those microbes. The acidophilic micro-organisms then take over and continue the reaction. The affinity for different pH can also be used to good effect to occlude spoilage organisms.

Inhibitors

Many chemical compounds can inhibit the growth and activity of micro-organisms. They do so by preventing metabolism, denaturation of the protein or by causing physical damage to the cell. The production of substrates as part of the metabolic reaction also acts to inhibit microbial action.

BENEFITS OF FERMENTATION

Fermenting fruits and vegetables can bring many benefits to people in developing countries. Fermented foods play an important role in providing food security, enhancing livelihoods and improving the nutrition and social well being of millions of people around the world, particularly the marginalised and vulnerable.

Improving food security

Eight hundred million people do not have enough food to eat. If we include those not free from hunger the figure rises to 1.2 billion people. This is one fifth of the World's population. A further two billion people are deficient in one or more micro-nutrients. In the seventies, food security was viewed mainly in terms of food supply at the global and national levels. Since then there has been a major shift in understanding of food security with more emphasis on access to food rather than purely on production.

The Food and Agriculture Organisation of the United Nations (FAO), amongst other influential organisations, has recognised that the problem of food security cannot be tackled in isolation. Moreover that it is an integral component of other development issues. FAO highlights the fact that the world food insecurity problem is a result of undemocratic and inequitable distribution of and access to resources rather than a problem of global food production.

Fermentation technologies play an important role in ensuring the food security of millions of people around the world, particularly marginalised and vulnerable groups. This is achieved through improved food preservation, increasing the range of raw materials that can be used to produce edible food products and removing anti-nutritional factors to make food safe to eat.

Fermented foods are popular throughout the world and the production of fermented food products is important in many countries in providing income and employment. In Africa, fermented cassava products (like *Gari* and *Fufu)* are a major component of the diet of more than 800 million people and in some parts of Africa it constitutes over 50%

of the diet. In Asia the preparation of fermented foods is a widespread tradition. *Kimchi* is the major food product of Korea. Soy sauce (a fermented legume product) is economically important from Indonesia to Japan. Over a billion litres are produced each year in Japan alone.

Over 2000 million litres are produced each year in Korea and over 150 million litres in Taiwan. *Miso* (a fermented legume product) is also very important in Asia with over 560,000 tons produced a year in Japan alone. In Latin America, fermented cereal products, alcoholic drinks and fermented milk products are three of the most important sectors of the economy.

REFERENCES

Beuchat, L.R. and A.D. Hocking, "Some considerations when analysing foods for the presence of xerophilic fungi", *J. Food Prot.*, 1990.

D'Aoust, J.-Y., "Effective enrichment-plating conditions for detection of Salmonella in foods", *J. Food Prot.*, 1984.

Haldane, J.B.S., "Sampling errors in the determination of bacterial or virus density by the dilution method". *J. Hygiene.*, 1939.

Hauschild, A.H.W. and R. Hilsheimer, "Enumeration of food-borne Clostridium perfringens in egg yolk-free tryptose-sulfite-cycloserine agar", *Appl. Microbiol.*, 1974.

7

Advanced Technologies in Microbiology

Robert Koch developed many classic microbial techniques. He used adaptations of the staining methods of Carl Weigert to begin the process of distinguishing microbes and identifying pathogens. His lab was the first to isolate a disease-causing organism.

A major contribution to bacterial techniques was the development of methods using solid medium for the cultivation of bacteria. Koch was convinced that microbes caused some diseases. However, to test this idea, he needed to isolate the causative agent. Almost all samples from diseased animals or any natural surface contained many different microbes and it was impossible to tell which one was the problem. A method was needed to separate these different bacteria.

The most common method of isolation was to continually dilute a sample in liquid broth in hopes that at high enough dilution, only one type of microbe would be found. A problem with this method is that only the most populous microbe would be isolated, but that might not be one causing the disease. There were other technical problems as well with such a liquid-based system, so a solid medium would seem to provide distinct advantages. Koch had tried gelatin for these experiments with unsatisfactory results. Building on the work of Brefeld and Schroeter, Koch used potato slices as a solid medium and observed that a boiled potato left in the open air would develop tiny circular raised spots.

Examination of these spots revealed they were made up of microorganisms and each spot had just one type of microbe in it. He realized that these colonies were pure cultures of bacteria and probably arose from a single species of microbe from the air that landed on the potato. By boiling a potato, slicing it with a hot knife and keeping it in a sterile container with a lid, Koch could keep the potato sterile. But if a sample from a disease animal was smeared across the potato, colonies arose, each being pure isolates from the animal. By then testing these isolates in animals, Koch was able to isolate the cause of anthrax, *Bacillus anthracis*.

Potatoes failed to support the growth of many microorganisms and Koch and his laboratory were constantly frustrated by the lack of a good solid medium. Walter Hesse joined Koch's laboratory to do studies on air quality, showing a remarkable attention to detail and patience in his work. His wife, Angelina Fannie Hesse along with raising their three sons, also would assist her husband with his research in the laboratory. Walter was attempting to do his air quality experiments using medium containing gelatin as the solidifying agent. In the summertime, temperatures would often rise above the melting point of gelatin, then used as the solidifying agent. In addition, microbes would often grow in the cultures that were capable of degrading gelatin and in both cases this would cause liquefaction of the medium, ruining the experiments. One day while eating lunch, the frustrated scientist asked Lina (as she was called) why her jellies and puddings stayed solid even in the hot summer temperatures. She told him about agar-agar, a heat resistant gelling agent that she had learned about while growing up in New York from a Dutch neighbour who had emigrated from Java.

Development of the new agent by Angelina and Walter led to a resounding success. Few microbes are able to degrade agar and it melts at 100 °C yet remains molten at temperatures above 45°C. This allows the mixing of the agar with heat-sensitive nutrients and microbes. After solidification, it will not melt until a temperature of 100 °C is again attained, facilitating the easy cultivation of pathogens. It can also be stored for long periods of time, allowing the cultivation of slow-growing microorganisms. Any type of broth can be mixed with agar, giving great flexibility in the kinds of medium that can be made. Thus, many more types of microbes could be cultivated.

Koch's laboratory also developed methods of pure culture maintenance and aseptic technique. Aseptic technique involves the manipulation of pure cultures in a manner that prevents their contamination by outside microorganisms. Equally important, aseptic technique prevents their spread into the environment. Remember that Koch was studying some of the most devastating microbial pathogens of the period, and their release could potentially cause disease in the scientists working on them. These procedures were also absolutely critical because they allowed careful study of pure microorganisms, making it possible to identify the role of each microbe in a given situation.

Another problem in the cultivation of microbes was solved by Julius Petri while working in Koch's laboratory. Solid medium was poured on glass plates and allowed to spread and harden. Once cooled it allowed a solid surface for streaking. However, creation of these plates required great care since exposure to the air (and the microbes in it) often lead to contamination. In addition, to prevent contamination of plates during incubation, a cumbersome bell jar was used. If one wanted to view samples, the plate had to be removed from the jar, further exposing it to unwanted microbes in the air. In 1887 Petri developed shallow glass dishes, with one having a slightly larger diameter than the other. Medium is poured into the smaller dish and the larger one serves as a cover. This simple device solved all of the above problems and took on the name of its inventor, the petri plate.

These same techniques are essential in studying all microorganisms. Collectively the above techniques have been used to isolate and identify thousands of different microorganisms. As a testament to the significance of their achievement, these techniques are practiced with remarkably little change in every laboratory that works with microorganisms today. Table 1 lists some of the early advances that helped to develop the practice of microbiology.

Table 1. The development of early techniques in microbiology

Year	Event
1664	Robert Hooke is the first to use a microscope to describe the fruiting structures of molds. He also coined the term cell when using a microscope to look at cork, as the dead plant material in cork reminded him of a jail cell.
1673	Anton van Leeuwenhoek, a Dutch tradesman and skilled lens maker, is the first to describe microbes in detail.
1872	Ferdinand Julius Cohn publishes landmark paper on bacteria and the cycling of elements. In it is an early classification scheme that uses the name *Bacillus*.
1872	Oscar Brefeld reports the growth of fungal colonies from single spores on gelatin and the German botanist Joseph Schroeter grows pigmented bacterial colonies on slices of potato.
1877	Robert Koch develops methods for staining bacteria, photographing, and preparing permanent visual records on slides.
1881	Koch develops solid culture media and the methods for obtaining pure cultures of bacteria.
1882	Angelina Fannie and Walther Hesse in Koch's laboratory develop the use of agar as a support medium for solid culture.
1884	Hans Christian Gram develops a dye system for identifying bacteria [the Gram stain].
1887	First report of the petri plate by Julius R. Petri.

1915 M. H. McCrady establishes a quantitative approach for analyzing water samples using the most probable number, multiple-tube fermentation test.

MICROSCOPY

Light Microscopy

In the 13th century, the British natural scientist Roger Bacon used segments of glass spheres as magnifying glasses and recommended them to partially sighted persons as spectacles.

The closer an object is to the eye, the more details can be distinguished until the limit is reached, below which the eye cannot depict the object clearly any more. This limit is 250 mm for the average grown-up. The measure for the ability to tell two points apart is called the power of resolution. It describes whether two adjoining points can still be perceived as separate. To examine objects below this limit of resolution, a magnifying glass is needed. Its magnification can be calculated in the following way:

$$V = 250\ [\text{mm}]\ /\ f\ [\text{mm}].$$

250 mm is the image distance of the human eye and f is the focal length of the lens. If the magnifying power of the lens is known and the focal length has to be calculated, than the formula has to be used as shown beneath:

$$f\ [\text{mm}] = 250\ [\text{mm}]\ /\ V$$

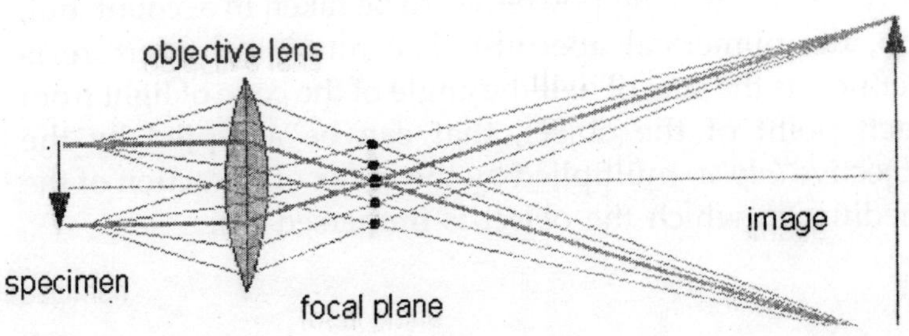

The picture to the above shows the way the fan of rays of an object that is collected by a single lens is imaged.

If you want to magnify an object even more, two lenses in tandem have to be used (objective and eyepiece or ocular). The resulting construct is a simple microscope. The objective magnifies the object or specimen (O) and a turned up real image (O′) is formed in front of the focal plane of the second lens. The eyepiece than forms an enlarged virtual image (O″), that can be seen as a turned up image at a distance of 250 mm. The magnification of a microscope is thus a product of

$$V_{objective} \times V_{ocular}$$

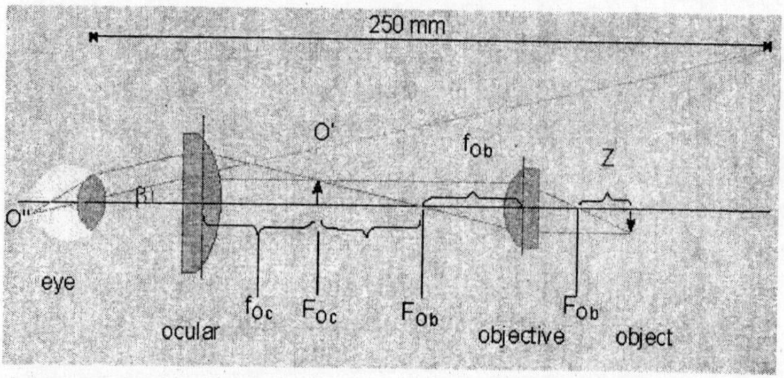

Light path in the microscope: F = Focal plane, O = Object (Specimen), Ob = Objective, Oc = ocular (eyepiece)

Nonetheless the power of resolution is given exclusively by the objective. To describe an objective not only the magnifying power has to be taken in account, but, too, the numerical aperture. The numerical aperture is defined as the sine of half the angle of the cone of light from each point of the object, that can be accepted by the objective *(alpha)* multiplied by the index of refraction of the medium in which the object is immersed (n):

$$A = n \sin alpha.$$

The medium is usually air with a refraction index of n = 1. *alpha* can never be bigger than 90° and thus the numerical aperture can never outgrow 1. Its largest actual size is 0.95, since the distance between objective and the surface of the cover glass cannot reach zero. The aperture of 0.95 corresponds to an angle *alpha* of roughly 72°. An increase of the numerical aperture can be achieved by the choice of a medium between objective and object with an index of refraction bigger than that of air. Special oil for immersion with an index of n = 1.515 has proved to be useful. Larger indexes of refraction do not make sense, because the index of refraction of the objective itself (n = 1.525) becomes limiting. Immersion oil can be used only with specially constructed immersion objectives. If *alpha* has the maximum of 67.5°, the aperture is accordingly 1.515 x 0.92 = 1.40. The degree of resolution (d) is set by the wavelength of light *(lambda)* and the numerical aperture (A_{obj}):

$$d = lambda \ / \ A_{obj}$$

If *lambda* is 550 nm (green light) the formula runs the following way:

$$d = 550 \text{ [nm]} \ / \ 2 \times 1.40 = 200 \text{ nm} = 0.2 \ \mu m$$

0.2 µm is the highest theoretical resolution that can be reached with a light microscope. A rough approximation shows that the power of resolution of a light microscope lies at about half the length of a lightwave if a good immersion objective is used. If the limit of resolution of a microscope is known, then the maximal useful magnification can be calculated. A magnification is called useful when two only just clear points are magnified so strongly that they are seen as separate unities by the human eye. At 250mm is the resolution of the human eye about 0.15 - 0.2 mm. The rule of thumb for a useful magnification is thus:

$$500 - 1000 \times A_{obj}.$$

An objective with an aperture of 1.4 has accordingly a maximal useful magnification of 1400fold.

Wave-Particle Nature of Light

Diffraction and Image Formation

We assume that the dual nature of light as both wave and particle is known and we will in the following deal with some fundamental principles of wave optics in order to be able to interpret microscopic image formation. Our main interest, nevertheless, is the understanding of modern techniques of light microscopy like phase-contrast microscopes, polarising microscopes and interference microscopes. First some terms and definitions:

Amplitude, frequency, wave length, interference, diffraction and phase are the most important parameters to describe a wave. The wavelength of visible light is between 400 and 800 nm. Interference is the mutual influence of two waves on each other, whereby the resulting crests may be either enhanced or flattened (enhancement of amplitude, reduction of amplitude). The extreme case is given by two waves that extinguish each other.

Diffraction is the partial deflection of a ray of light at the corners of opaque objects. For understanding it is useful to cast a look at phenomenons of diffraction at simple gap- or hole-shaped masks. If a screen is placed behind such a mask and light is sent through, the pattern of diffraction can be seen.

It consists of a regular pattern of points or lines that fades away towards the margins. The shape of the pattern is dependent on the kind of mask and the object used (for example a microscopic preparation). But the pattern is always grouped around a central axis. If a mask with two openings is used, the light is diffracted at the rim of both openings. The diffraction images of the two rays overlie each other and regular enhancements and reductions can be seen in the resulting pattern.

To get a magnified image, the first additional maxima of the diffraction image have to be included. The results in the pictures above have already shown that the distances of the points in a diffraction image are reversed proportional to the corresponding distances in the object. To gain a real image from a diffraction image a collecting lens (the objective of the microscope is one) is needed. This lens has to have a big enough aperture to collect the intensity maxima (-I, +I). The real image is formed by both refracted and non-diffracted light.

Diffraction phenomenon at apertures. The closer two openings (or diaphragms) that diffract light are, the larger is the angle of the diffraction's first intensity maxima and the larger has the aperture of the collecting lens to be.

Magnifications as those that have been described can, too, be achieved with radiation of other natures. The most important additional device for biologists besides the light microscope is the electron microscope, where the diffraction of a ray of electrons is used. Another radiation to be mentioned is the short-waved Röntgen-radiation (X ray-radiation), that has proven to be of value for the elucidation of molecular structures.

Since no collecting lenses for Röntgen-radiation exist, the information contained in the diffraction pattern has to be calculated mathematically in order to get to know the structure of the molecule.

PHASE-CONTRAST AND DARK FIELD MICROSCOPY

A main obstacle in the microscopy of biological objects is their poor contrast.

Light-absorbing parts of a preparation weaken the amplitude of the light waves that pass through them. It is thus also spoken of amplitude preparations. The change of stronger and feebler light is perceived by the eye as a difference in brightness.

The invisible parts of the preparation are went through by the light without a change of amplitude, but the phase of the light may be changed depending on the consistency of the material. This change is due to the altered speed of the lightwaves. Differences in phase can be perceived neither by the eye nor by a photographical film.

FLUORESCENCE MICROSCOPY

Fluorescence microscopy is based on the fact that some molecules emit part of the light absorbed by them as longer waves. A well-known example is the red fluorescence of chlorophyll. Since the Austrian teacher M. Haitinger made his examinations in the thirties of this 20th century it is known that a number of so-called fluorochromes exist, with which microscopic preparations can be stained so that they emit fluorescence indirectly.

In the following decades, a whole range of so-called vital dyes were detected or developed, that -used in low concentrations- could mark specific parts of living cells or of tissues. They enabled researchers to follow the solute transport within cells and tissues or to ascertain the pH of special compartments, like for example the vacuole.

Since roughly twenty years fluorescence microscopy is flowering again, on one hand, because the range of fluorochromes became much broader and on the other hand, because completely new approaches like indirect fluorescence could be used successfully. Additionally the construction of microscopes was improved and highly efficient masks were developed.

There are two ways in which fluorescence microscopes can be constructed: as epifluorescence- and as transmission microscopes. The second is the older way of construction. Three components are needed:

1. A strong source of light, that emits mainly short lightwaves. Mercury high pressure lights have proven useful.
2. *The first barrier filter*: This filter helps to shut off all radiation other than the one that activates the specific dye. It is placed behind the light source within the light cone. Besides it is advantageous to work with a dark field condenser.
3. *Second barrier filter*: This filter is brought into the light cone between objective and eyepiece. It lets through only long wavelengths, that are caused by emission of the preparation (so-called "secondary radiation").

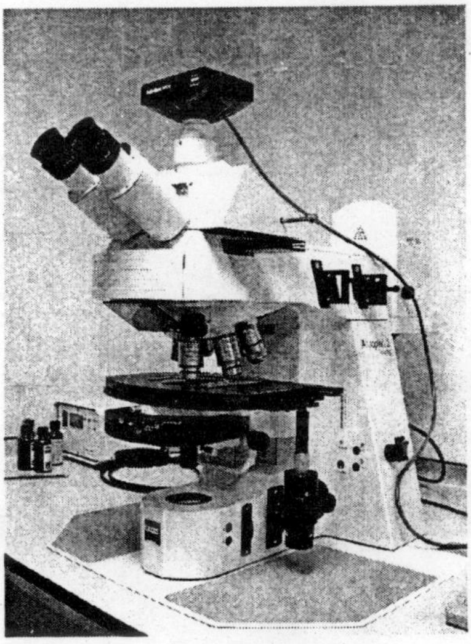

Fluorescence microscope

In the last years the **epifluorescence microscope** has replaced the transmission fluorescence microscope more and more. But the transmission fluorescence microscope is still better suited to only weakly magnifying objectives (2.5

x, 6.3 x). With the epifluorescence the objective is also a condenser and the stronger it is the more intensive radiation can be used. The heart of epifluorescence is a construction within the light cone between objective and eyepiece that serves to feed activating radiation into the system and is constructed of first barrier filter, beam-splitting mirror and second barrier filter.

POLARIZATION MICROSCOPY

Light waves do usually radiate into all directions. Suitable polarization masks allow the elimination of certain planes of polarization, whereby linear polarized light is gained.

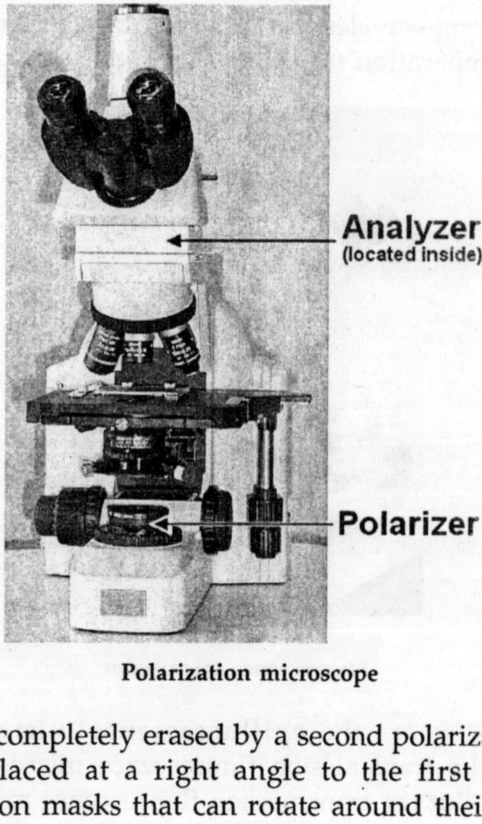

Polarization microscope

It can be completely erased by a second polarization mask that is placed at a right angle to the first one. Such polarization masks that can rotate around their own axis can be placed in the light cone of a microscope. One beneath the condenser and the other above the objective.

The use of a thus constructed polarization microscope is only sensible, if preparations with polarizing properties are examined. This is the case, if the preparation consists of units that are oriented in a certain way like molecules or atoms (double refracting crystals). This type of microscope is thus used mainly in mineralogy since crystals are *per definitionem* built in a regular pattern. By the use of polarized light crystal axis and three-dimensional can pattern be ascertained exactly. The practical application in biology is somewhat limited, but the structure of starch grains for example or the orientation of cellulose fibrils within the vegetable cell wall or the orientation of rod-shaped viruses (tobacco mosaic virus) can be determined.

Based on polarization microscopy the French physicist G. Nomarski developed the interference contrast microscope (also called differential interference contrast, DIC) in Paris during the middle of the 1950s. For this type of microscopy two additional Wollaston-prisms are needed. A Wollaston-prism is composed of two cemented calcium fluoride wedges. A polarized ray of light is split into two beams at the cemented plane that are vertical towards each other .

MICROPHOTOGRAPHY

The development of microscopy created a need for the documentation of the images. The classical approach that is practized by every biology student is drawing. A help was the construction of a drawing device invented by E. Aabbe, that is placed upon the eyepiece and projects the image *via* prism and mirror onto a drawing pad. An alternative came into being with microphotography, that has the disadvantage of a very small depth of field and captures consequently only single planes of the object. This is the reason why it is until today viewed rather critically by some biologists. But it has nevertheless been established as an independent method.

Numerous images on our web pages show that an exact image of what can be seen is given by a photo. If it is necessary, several images of different planes of the object can be taken. The important advantages compared to drawings are the speed of documentation and the exclusion of subjective perceptions (freedom of the artist), whereby a lot of damage can be caused. One and the same object can be studied in parallel with with different microscopic methods, corresponding pictures can be compared in order to determine their orientation. The film material of today brings astonishing results even at the very poor light conditions of fluorescence microscopy.

The possibility of cinematography has already been mentioned shortly. Many documentaries of cellular motion, of cells, their coupling etc. exist. Most of the documentaries produced in Germany were done in co-operation with the Institut für den Wissenschaftlichen Film in Göttingen. It is possible for other institutes and other scientific institutions to borrow them.

Video cameras, image intensification and computer-controlled image-storing and -interpretation do also belong to modern microscopy. They have first been used at the beginning of the fifties, but the important developments started during the last years of the seventies. Two processes especially that have been developed by R. D. Allen, N. Stromgren-Allen and J. F. Travis have to be mentioned:

— AVEC-DIC (*Allen video-enhanced contrast-differential interference contrast*) and

— AVEC-POL (*Allen video-enhanced contrast-polarization microscopy*)

Both methods are based on the fact that certain video cameras (the right choice of the model is decisive) can process differences in brightness on a completely different scale than the human eye or a photographic film. Depending on the study or the object can different adjustments of the camera be chosen. E. Abbe defined the resolution limit of the light microscope as that distance between two adjoining points, at which they can only just be perceived as different unities. The limiting factor is the wavelength of the used light. This resolution limit does not mean that smaller structures stay invisible. Fluorochromization, for example, allows single molecular complexes to be seen, if the distance between molecules of the same kind is above the resolution limit of the microscope.

Instead of fluorochromization contrast-enhancement by AVEC-DIC can be used and although the contrast produced by microtubuli is not big enough to be perceived by the human eye it can be recorded and identified by a camera.

Image-storing and interpretation allow the emphasizing of movable particles. The appropriate computer program subtracts the content of one image from that recorded seconds or minutes afterwards of the same spot. All immovable structures vanish and only the moveable ones are depicted.

Cameras with image intensification are used in fluorescence microscopy to document fluorochromes with a fluorescence that is far below that perceivable by the human eye. Again it is possible to use the already mentioned method of vital staining, because far more specific sondes are known today. Since their concentration is smaller than that used in the forties or fifties the cells are damaged less.

While in the nowadays common fluorescence microscopy photons (h*ny*) are emitted and recorded after the illumination of the object, a photoelectron microscope allows the tracing of secondary electrons, that are set free after exposure to UV radiation. This microscope has electronic optics and the resolution is higher than that of fluorescence microscopes. The technique has only recently been developed, but first results with biological objects exist already. They suggest that the method will be widespread within short time.

ELECTRON MICROSCOPY

In 1924 L. de Brogile discovered the wave-character of electron rays thus giving the prerequisite for the construction of the electron microscope. The prototype was built by M. Knoll and E. Rruska. One of the first biological objects depicted was the tobacco mosaic virus (TMV).

Transmission Electron Microscope (TEM)

The conventional electron microscopy is nowadays called TEM (transmission electron microscopy). We will therefore start with its construction. The ray of electrons is produced by a pin-shaped cathode heated up by current. The electrons are vacuumed up by a high voltage at the anode. The acceleration voltage is between 50 and 150 kV. The higher it is, the shorter are the electron waves and the higher is the power of resolution. But this factor is hardly ever limiting.

The power of resolution of electron microscopy is usually restrained by the quality of the lens-systems and especially by the technique with which the preparation has been achieved. Modern gadgets have powers of resolution that range from 0,2 - 0,3 nm. The useful resolution is therefore around 300,000 x.

The accelerated ray of electrons passes a drill-hole at the bottom of the anode. Its following way is analogous to that of a ray of light in a light microscope. The lens-systems consist of electronic coils generating an electromagnetic field. The ray is first focused by a condenser. It then passes through the object, where it is partially deflected. The degree of deflection depends on the electron density of the object.

The greater the mass of the atoms, the greater is the degree of deflection. Biological objects have only weak contrasts since they consist mainly of atoms with low atomic numbers (C, H, N, O). Consequently it is necessary to treat the preparations with special contrast enhancing chemicals (heavy metals) to get at least some contrast. Additionally they are not to be thicker than 100 nm, because the temperature is raising due to electron absorption. This again can lead to destruction of the preparation. It is generally impossible to examine living objects.

After passing the object the scattered electrons are collected by an objective. Thereby an image is formed, that is subsequently enlarged by an additional lens-system (called projective with electron microscopes). The thus formed image is made visible on a fluorescent screen or it is documented on photographic material. Photos taken with electron microscopes are always black and white. The degree of darkness corresponds to the electron density (= differences in atom masses) of the candled preparation.

Scanning Electron Microscope (SEM)

The path of the electron beam within the scanning electron microscope differs from that of the TEM. The technology used is based on television techniques. The method is suitable for the depiction of preparations with conductive surfaces. Biological objects have thus to be made conductive by coating with a thin layer of heavy metal (usually gold is taken).

The power of resolution is normally smaller than in transmission electron microscopes, but the depth of focus is several orders of magnitude greater. Scanning electron microscopy is therefore also well-suited for very low magnifications. Numerous examples will be given in the following.

The surface of the object is scanned with the electron beam point by point whereby secondary electrons are set free. The intensity of this secondary radiation is dependent on the angle of inclination of the object's surface. The secondary electrons are collected by a detector that sits at an angle at the side above the object. The signal is then enhanced electronically.

The properties of the light microscope as opposed to that of transmission and scanning electron microscopes are collected in a table.

The high voltage electron microscope: it operates with an accelerating voltage of 700 - 3000 kV. Its power of resolution is greater, the preparation can be thicker, the strain on the preparation is smaller. But the enormous technical expenditure is disadvantageous. Only few gadgets exist. New results concerning botany have not been gained.

The scanning transmission electron microscope (STEM): In this development of the SEM do the electrons pass through the preparation and the secondary radiation thus generated is used for image formation. Here, too, the expenditure is large, but it is still worthwhile, since large molecules like nucleic acids or proteins or molecular complexes like viruses can be depicted much better and gentler than with the TEM. No news for botany, though.

The interpretation of images gained with electron microscopy is increasingly done with computerized interpretation programs. But they are usually only suitable for the reconstruction of regularly recurring patterns and these, again, are found more often on a molecular level than on a cellular one.

ENRICHMENT TECHNIQUES

Winogradsky Column

This classic enrichment technique was devised by the early soil microbiologist Winogradsky. The method selects for organisms that:

1. Use light as an energy source (photosynthetic).
2. Grow in a range of oxygen conditions (from strict anaerobes to obligate aerobes).
3. Use sulfur-containing molecules (such as H_2S, which is a gas that smells like rotten eggs) as electron donors in photosynthesis.

Method

1. *The mud.* First, collect about half a bucket of mud from a swamp or marsh. A salt marsh at low tide is usually a good source, soft, smelly mud is best. Take your sample from the surface layer, where there is enough light for photosynthesis.
2. *The container.* Any tallish, transparent glass or plastic cylinder will work. A good container can be obtained by simply cutting off the top, narrow part of a 1-liter plastic soda bottle. In the lab, glass or plastic chromatography columns or graduated cylinders can also be used.

Preparation

1. After rinsing the container, place some bits of paper towel and the yolk of a hard-boiled egg in the bottom. These will provide nutrients and, in particular, a source of reduced sulfur (i.e., hydrogen sulfide).
2. Pour in the mud to a level just below the top of the container. Push the mud down until all air bubbles are gone. Pour a bit of water (~2 cm) into the top and cover the open end to prevent the mud from drying out.
3. Put the cylinder in a place where it won't be disturbed and illuminate it from one side by using an ordinary incandescent lamp or simply placing it in a windowsill.

Results

After a few days to a week or so, colored areas will start

to appear on the illuminated side. These are colonies of anaerobic and microaerophilic photosynthetic sulfur-using bacteria. Over a longer period, yellowish deposits of elemental sulfur may accumulate.

Implications

The microbes that appear so colorfully in this enrichment method are quite interesting and may resemble microbial communities present in the early period of life on earth. The atmosphere of the early earth was quite different from the one we have today. For one thing, it had no free oxygen. However, a number of other gases, such as methane and hydrogen sulfide, were present. Hydrogen sulfide is chemically similar to water. Water has the chemical formula H_2O, and hydrogen sulfide has the chemical formula H_2S. As you can see, the only difference is that water has an oxygen atom in the place where hydrogen sulfide has a sulfur atom.

Modern plants use the energy of the sun to manufacture glucose from water and carbon dioxide, in the process giving off oxygen derived from the water. This is the origin of oxygen in our modern atmosphere. More specifically, the oxygen comes from the water molecule, which the plant cells take apart using the energy of the sun. In the early days on earth, the first microbes to use the sun's energy did not give off oxygen.

Some of them used hydrogen sulfide and carbon dioxide to make simple compounds, in the process giving off pure sulfur. The great sulfur deposits in places such as Louisiana come from the ancient activities of these primitive bacteria. These bacteria still exist today and are found in sediments and stagnant waters where there is: (i) essentially no oxygen, (ii) some light, and (iii) a source of hydrogen sulfide (usually a breakdown product of the activities of other microorganisms). In setting up the Winogradsky column you have recreated some of the conditions of the early earth and thereby enriched for very interesting, early types of bacteria.

Baiting Traps for Microbes

In this method we enrich for certain protists, ciliates and flagellates, which feed on bacteria and decomposing matter.

Materials

Start with a film canister (the black cylindrical plastic container that rolls of film come in, any film developing shop will be glad to give you some). In addition, you will need a small amount of material to serve as a screen, a bit of fine silk or nylon from a discarded stocking or panty hose will work. You can also use a nylon material called Nytex, which is used for making plankton nets.

Making the trap

Using a sharp knife or razor, cut a small square "window" in the side of the canister. Then glue a square of the silk or nylon material to the edges of the window, forming a screen. Any nonwater-soluble glue should work (e.g., crazy glue, nail polish).

Using the trap

Remove the lid and put a bit of canned tuna fish (water based, not oil based) in the canister. Put the lid back on and submerge the canister in an aquatic habitat where there is a lot of organic matter and protozoa should be numerous. Make sure to let any bubbles of air in the canister escape. Since the canister floats, you may want to tie a weight to it or simply put a stone inside to weigh it down.

After a period of a day or more, retrieve the canister. With luck, it will contain a dense population of protists, which have been attracted by the tuna and have multiplied inside the trap, feeding on bacteria and perhaps directly on the decomposing tuna fish. The purpose of the screen is to screen out larger organisms, such as small fish and various invertebrates that would also be attracted by the tuna fish. You might like to try some variations: other containers, other baits.

For example, an early pioneer, Pringsheim, used cheese to attract and enrich for flagellates that could live on small-chain fatty acids ("acetate flagellates," such as Euglena and Chlamydomonas and their relatives). You might have other ideas. Tuna fish generally works well.

Enriching for Soil Amoebas

The soil is full of bacteria, and it is also full of things that eat bacteria. In this method, we enrich for one component of these bacteria, the small soil amoebas.

Materials

1. Petri dishes containing water agar with no added nutrients.
2. A culture of a suitable food bacterium, such as Enterobacter aerogenes.
3. A bit of soil.

Method

Place a small amount of soil mixed with water on the agar gel surface in the center of the petri dish. Using a bacteriological loop or a dropper, carefully make a ring of food bacteria around the bit of soil. Between the soil and the surrounding bacterial ring there should be bare agar surface, at no point should the soil touch the ring.

Results

The small amoebas in the bit of soil will crawl away from

the soil over the agar surface. When they reach the ring of food bacteria they will start to consume them and will multiply. Using the microscope you will be able to see them clearly after a few days (low power, 100x total magnification). Since soil is a complex habitat with a rich and diverse biological component, you will often see other things besides soil amoebas coming out in these plates, such as round worms (nematodes) and fungi.

Enrichments for Extremophiles

What we consider an extreme condition may be normal for certain organisms. Here are some possible enrichments:

— *Thermophiles*. Inoculate some nutrient broth with a bit of soil or other source (e.g., water from the hot water tap or pasteurized milk), then incubate at a high temperature. Organisms that will grow at 55oC or above are called thermophiles. If the broth becomes cloudy and smelly from bacterial growth, you may wish to try to isolate a species by making a dilution series. Add 1 ml of the primary enrichment to 9 ml of sterile broth, mix by shaking gently. Then add 1 ml of the first dilution to 9 ml of sterile broth, mix, and so on for several dilutions. With luck, you may obtain a pure or nearly pure culture of the most abundant species in a high dilution.

— *Acidophiles*. Adjust the pH of the nutrient broth to pH_3 and inoculate with soil. After growth appears, you may try isolation by serial dilution, as above. In addition to soil, you might try homemade vinegar, pickles, or sauerkraut.

— *Halophiles*. NaCl is added to nutrient broth to give a final concentration of 8% (8 grams of NaCl per 100 ml of nutrient broth) or more. In addition to soil, seawater or saline inland waters would be good sources, and you might try pickles, salted meats, or sauerkraut.

Oil Degraders

Hydrocarbons are produced naturally in marshes and other wetlands by microorganisms. In addition, oil spilled from tankers and pipelines and runoff from highways and streets cause oil contamination of soil and water. Of particular concern are polyaromatic compounds, and especially the chlorinated ones (e.g., polychlorinated biphenyls or PCBs), as many of these are potentially hazardous to humans. Naturally occurring microbes exist that will break these compounds down, and some of these are now used for bioremediation of oil spills.

Method

A suitable inoculum is any soil or natural water, particularly from a marsh or a site frequently impacted by oil spills (e.g., soil around a gas station). Here are suggestions for two kinds of basal media that can be used:

1. Standard nutrient broth medium to which hydrocarbon is added.
2. Natural water with added nitrogen, phosphorus compounds, iron, and other trace elements. A convenient source is commercially available plant food, such as Miracle-Gro.

To either of the basal media, add a hydrocarbon. Powdered mothballs (either naphthalene or dichlorobenzene) are a convenient source of aromatic hydrocarbons and form a good enrichment. Similarly, waste oil, which supports growth on an aliphatic compound, works well too.

Results

After a week or more, the medium will become cloudy from bacterial growth and the hydrocarbon will start to disappear. A subsequent transfer of the established culture into fresh inoculum at this time may be helpful. Once the oil degrader is established in culture, detection of the microorganism may be confirmed through the elimination of the hydrocarbon, as well as microbial growth. One can also confirm the presence of oil degrader respiration, as one would heterotrophic organisms, by calculating the change in Biological Oxygen Demand or oxygen consumption.

Cellulose Digesters

Cellulose, the most common natural polymer, is produced in enormous quantities by plants and then degraded by fungi and bacteria. Animals such as cows and termites have symbiotic microbes in their intestines that can digest this refractory polysaccharide.

Method

To enrich for cellulose degraders, first find a plausible inoculum source, such as rotting wood, decaying leaves, cow cud, or termite gut. Place a small amount of this inoculum on carboxymethyl cellulose (CMC) agar plates. After a week or more, there should be a significant amount of growth on these plates. Pick a small area of growth and restreak this onto another CMC agar plate. Incubate these plates for approximately a week. To determine whether cellulose degradation has occurred, stain the plates with Congo red. This is done by pouring a solution of Congo red onto the plates and letting them sit for 10 minutes.

After 10 minutes, pour this solution into a waste beaker (to be autoclaved or sterilized using bleach solution). Then destain by pouring a solution of 1 N NaCl onto the plates. Let this sit for 10 minutes and then pour the solution into the waste beaker. The Congo red binds to cellulose. Therefore, degradation of the cellulose should be evident based on "cleared" rings.

Agar Digesters

Agar is widely used as an agent to solidify media for

microbiological studies. This polysaccharide is produced by red algae and is ultimately degraded by fungi and bacteria.

Method

To enrich for agar degraders, first find a plausible inoculum source, such as salt marsh mud or soil. Place a small amount of this inoculum on water agar plates. After a week or more, there should be a significant amount of growth on these plates. Check the plates for areas of indentation in the media. This may indicate bacteria are present that are capable of degrading the agar. Pick a small area of growth from one of these indentations and restreak onto another water agar plate. Incubate these plates for approximately a week or two. Deep "pits" in the plate indicate that you have indeed isolated an agar digester.

Starch Digesters

Starch is a polymer of α-1,4-linked glucose that is naturally produced by plants (i.e., leaves and seeds). Many fungi and bacteria produce amylase, which is used to digest this polysaccharide. A common bacteria that you may isolate which can hydrolyze starch is the Bacillus spp.

Method

To enrich for starch degraders, first find a plausible inoculum source, such as rotting wood or leaves. Place a small amount of this inoculum on starch agar plates. After a week or more, there should be a significant amount of growth on these plates. Pick a small area of growth and restreak this onto another starch agar plate. Incubate these plates for approximately a week.

To determine whether starch degradation has occurred, stain the plates with iodinepotassium iodide. This is done by pouring a solution of Gram's iodine onto the plates to flood them and letting them sit for 1 minute. After 5 minutes, pour this solution into a waste beaker. Degradation of the starch should be evident based on "cleared" rings. Their presence can be established by transferring several likely colonies (i.e., white, waxy texture) to separate plates. After a short period of incubation, confirmation of a positive Gram test and positive spore stain will verify the presence of Bacillus.

Nitrogen Fixers

Diatomic nitrogen is the most abundant element in the atmosphere, but few organisms can use this resource directly. Nitrogen is often the limiting nutrient in natural ecosystems, as well as in agriculture. Thus, the nitrogen fixers, bacteria that can convert atmospheric nitrogen to biologically available forms, can have great ecological significance. Winogradsky was the first to isolate nitrogen-fixing bacteria from soil.

Method

The basic idea of this enrichment is to create an environment containing everything except nitrogen, so that organisms that can use atmospheric nitrogen will be favored. The method described in the isolation exercise for Azotobacter is inspired by the classic approach developed by Martinus Beijerinck and involves incubating a paste made of soil and calcium carbonate, with added mannitol as both the carbon and the energy source, and dibasic phosphate as both the buffer and the phosphate source. As you establish an enrichment culture for these microorganisms, it is important to note that nitrogen fixers prefer a slightly alkaline pH. Both Rhizobium and Azotobacter species are among the relatively few bacteria that can metabolize mannitol, which is produced by plants.

BACTERIAL STAINING TECHNIQUES

Bacteria are too small and too transparent to be well

described using light microscopy and a wet mount. To make them more visible by imparting contrast, they are stained. Most cellular staining that takes place falls into three categories.

Simple stains color bacteria to make them easily visible under the light microscope. A simple stain is a basic (positively charged) dye that interacts with the negatively charged macromolecules of the cell (membrane, proteins and nucleic acids) through electrostatic attraction. Methylene blue, Safranin and Giemsa are examples of simple stains. Simple stains permit the description of the size, shape and arrangement of cells on a slide or within a clinical sample.

Differential stains impart color to some but not all microbes. Those that bind and retain the stain are termed positive, those that do not are termed negative. In a differential staining procedure, the cells which do not retain stain become transparent after the differential staining procedure, and so are made visible with a simple staining procedure.

Thus differential staining procedures have two components: a differential stain, followed by a simple stain. The dyes used in each are different colors, so that a positive cell will be one color and a negative cell another. Thus, in addition to the descriptions possible with a simple stain (size, shape and arrangements), differential stains also permit classification as either positive (retaining the stain) or negative (only stained by the counter stain.) The Gram stain and the Acid-fast stain are two of these procedures.

REFERENCES

Booth, C., *Methods in Microbiology*, vol. 4. Academic Press, New York, 1971.

Dring, D.M., "Techniques for microscopic preparation", *In* C. Booth (ed.), *Methods in Microbiology*, vol. 4. Academic Press, New York, 1971.

McGinnis, M.R., *Laboratory Handbook of Medical Mycology*, Academic Press, New York, 1980.

8

Laboratory Techniques and Microbial Diagnosis

A number of techniques are routine in microbiology laboratories that enable microorganisms to be cultured, examined and identified.

An indispensable tool in any microbiology laboratory is the inoculating loop. The loop is a piece of wire that is looped at one end. By heating up the loop in an open flame, the loop can be sterilized before and after working with bacteria. Thus, contamination of the bacterial sample is minimized. The inoculating loop is part of what is known as aseptic (or sterile) technique.

Another staple piece of equipment is called a petri plate. A petri plate is a sterile plastic dish with a lid that is used as a receptacle for solid growth media.

BEGINNING OF A MICROBIOLOGY EXPERIMENT

This experiment does not require sterile media, but keep things as clean as possible.

The first microbiologists did not have culture tubes, fancy equipment, or special supplies. Just like them, we can accomplish much with the same kitchen foods they used. If you have culture tubes, use them, but otherwise you can conduct this experiment using a drinking glass or small bottles or any other useful container. One advantage of culture tubes is that it costs less to fill them, but the materials for this experiment are cheap.

Our first experiment is very inexpensive. We will try to culture (grow) baker's yeast (from the grocery store) on various carbohydrates learn what it needs to grow. Before a microbiologist can do much with a microbe, he must find a way to grow it. As we work with baker's yeast and then other microbes, you will learn what is needed to grow microbes and what containers will be useful. Thus you can begin building up supplies and equipment for your home microbiolgy lab.

Experiment 1. Growing a Baker's Yeast Culture

Growing a yeast culture is simple and easy, yet it introduces many concepts you need to understand if you want to conduct a serious microbiology project. This experiment will demostrate some of the biochemical principles used to identify bacteria and yeast species. You can find everything you need at home or in the grocery store and that is the first reason we are beginning with bakers yeast, *Saccharomyces cerevisiae*, but another reason is that since we eat yeast, we know it is safe for beginners to work with. This simple experiment can be expanded into many more complex experiments and become an outstanding science project.

The standard steps for growing a microbe are

— study the literature (books, journals, and web) to learn what food, light, temperature, and other conditions it needs for growth, or just make a guess from your knowledge of its habitat.

— prepare the culture medium (the liquid or agar mixture for growing it)

— put the medium in tubes, plates, or other container

— sterilize or pasteurize the containers of medium

— store the sterile containers of medium under suitable conditions

— inoculate the sterile medium with the microbe
— incubate the culture containers under suitable conditions: temperature, light if needed
— observe the growth, production of gas, and other details
— record results for future use and to plan future experiments.

Preparation of the medium

Medium is the name for the food we grow microbes in. The medium must contain everything our yeast needs.

— First we need to figure out what this yeast needs. Bakers use flour, table sugar, and water to make bread. In nature, yeasts are found in flower nectars, in slime fluxes (sap running out of tree wounds; such sap is alive with yeasts and insects that like sugars and yeasts).
— Recall that living organisms need a carbon source, an energy source, a nitrogen source, minerals, and may need vitamins. Carbon and energy source is often a single compound such as glucose (corn syrup). If in doubt, always try corn syrup.
— We will try corn syrup and water in this experiment. We will hope the water has enough minerals to supply the yeast's needs. After all, home bakers do not add any minerals.
— This mixture may not have enough nitrogen for the yeast to grow. If the yeast does not grow we could try meat broth, peptone, or another source of amino acids. The dry power yeast probably contains some items the yeast needs. Very likely the manufactuer adds them to insure the home baker is successful. However, we are not using flour and flour does contain proteins, minerals, vitamins, and starch.

— It tapwater does not have enough minerals, we could try water from a mud puddle or soil water made by mixing soil with water and using the clear supernatant water after the mud has settled. Such soil extract will contain lots of bacteria and should be sterilized first.

— How much corn syrup shall we use? Microbiologists commonly use 10 grams of the carbon and energy source per liter (1.1 quart). Corn syrup is mostly glucose which is a sugar most microbes can use. Recalling that a liter is 1000 mL or 1000 cubic centimeters, we will add 10 mL of corn syrup to a liter of water. Your ruler is marked in centimeters. So we want to add a slug of corn syrup about 1 cm x 1 cm x 10 cm long. You can estimate this volume easily and you need not be exact. A level tablespoon of corn syrup is probably about 10 to 13 grams of glucose content.

— Corn syrup and water may be an adequate medium, but dry yeast has some minerals and perhaps some nitrogen matter; lets try it and see if the yeast will grow. Read the label on your syrup. Much syrup sold today contains corn syrup and high fructose syrup. Fructose is sweeter than glucose and some microbes can use both sugars. Later we will study microbes which can't grow in the presence of sugars and other organic matter.

Tube your medium and sterilize it

— Normally we put our medium in culture tubes, plug them with cotton, and then sterilize them. We will omit the sterilization step for this experiment. In this experiment we are using so much yeast that the yeast will overwhelm the contaminants. In normal bread making we do everything so fast the contaminants do not have time to grow and poison us. Besides the yeast probably makes substances which interfere with the growth of other organisms.

— It would be nice to use a gas collection tube in this experiment. If you can find a small bottle or small culture tube, fill it with the syrup-water- yeast mixture and invert it in one of your culture containers with the medium. If the gas tube contains a bubble try again. The gas tube will collect gas if your growing yeast makes any gas. If you know some chemistry, you could try to identify any gas produced. [Hydrogen explodes, methane explodes, carbon dioxide and nitrogen extinguish flames by failure to support combustion].

Inoculate your culture(s).

— Inoculate means to introduce a microbe into a tube of medium.

— You can inoculate your tube or tubes with a few bits of the dry baking yeast you got at the grocery store. Save the remainder of the package for future experiments.

— A better way to inoculate tubes is to use the same amount of inoculum in each tube. Therefore, you dissolve a little dry yeast in water and add one drop of the suspension to each tube of medium. Limit the amount of yeast added so the culture is not cloudy. We like for cultures to be as clear as possible at the beginning so it will be easier to see the cloudiness as the yeast cells begin to grow.

— This will not be a pure culture of yeast because the water, containers, and your fingers carried various fungi and bacteria. Since yeast grows rapidly and outnumbers the contaminants, the contaminants will not have much effect. That is why bakers do not sterilize their medium. Bakers just try to keep things clean as possible. However, due to the contaminants the bread will become moldy after a few days. Commercial bread contains calcium propionate to retard growth of molds. Look at any bread label; what is the additive (hint it gives cheddar type cheeses some of their sharp flavor).

— Do not inoculate all your tubes, save some as controls. Incubate the controls beside the experimental tubes to determine whether control tubes remain unchanged. Since we did not sterilize our medium, the control tubes will likely get turbid soon and eventually rot.
— Any leftover syrup-water will rot. Therefore, you might as well discard it right away.

Incubate your yeast cultures

— Place them in any warm place to grow.
— If you wish place some tubes at different temperatures, such as, in refrigerator, hottest place you can find, and other temperatures in between. Suggested temperatures are 4C (refrigerator), 25C (room temperature), 35C (very hot room), 15C (very cold room)
— Mark your tubes so no adult or child drinks them. The sugar water is harmless, but the contaminating microbes may not be safe.

Record the results

Examine your tubes for growth at daily or 12 hour intervals. Discard any tubes that get moldy, wash them carefully so you do not inhale any mold spores. If any tube breaks, do not get cut by the contaminanted glass. If you get cut, wash the wound with clean soap and water and flood with alcohol.

— Did the clear medium become cloudy (evidence of multiplying cells)?
— Did any odor develop (Never sniff moldy tubes)?
— Did any sediment develop in the tubes; compared to the uninoculated tubes?

— Did any gas collect in the filled inverted gas tube inside your culture?
— Did growth occur differently at different temperatures?
— Did any growth occur at the hottest temperature. If not move that tube to the temperature that gave the most growth. Does it then grow? Some microbes die at high temperatures. (Of course very high temperatures will kill any microbe, but the killing temperature varies. If spores are produced, they survive much higher temperatures.
— Write down any ideas you have for improving your next experiment.
— As time goes on, growth will slow down and eventually all the cells will die.
— Does this yeast make spores? How can you determine that?

If you wish, you could perform the above experiment using a bit of garden soil (or slime flux) as the inoculum instead of bakers yeast. Since garden soil contains dozens of different species of bacteria and fungi, the results will be different and perhaps useless because soil usually contains bacteria capable of growing at most temperatures and on most foods.

Experiment 2. Growth of Bakers Yeast on Other Sugars

Biochemists have discovered dozens of sugars and other carbohydrates in nature. Various species of bacteria and yeasts differ in their ability to grow on these carbohydrates. The ability of a strain of bacteria or yeast to grow on certain sugars is one of the main methods for indentifying them to Genus and species. Since fungi have huge genomes and the genetic codes to make enzymes to digest most carbohydrates, sugar digestion is not as useful in for identifying fungi.

— You can repeat Experiment-1 using different carbohydrates. If the yeast grew on corn syrup water medium, try making media from water and each carbohydrate you can find.

— Starch and table sugar are the other two carbohydrates found in most kitchens. Some people may have maltose. Can you make some maltose by using sprouting barley which contains the enzyme maltase?

— Table sugar (sucrose) is a disaccharide made by plants (maple, beet, cane, and others). Plants make sucrose by joining one molecule of glucose to one molecule of fructose. Microbes can't use sucrose unless they have an enzyme capable of splitting sucrose apart to give glucose and fructose.

— Starch is a long molecule composed of hundreds of glucose molecules hooked together to make a long chain-like polymer. Microbes can't use starch unless they have enzymes to break it apart into separated glucose molecules. You probably should boil the starch in water before you use. It will still be long chains, but the boiled solution will be cloudy instead of white.

— Maltose is a disaccharide consisting of two molecules of glucose joined together covalently.

— Lactose is a disaccharide consisting of one molecule of glucose and one molecule of galactose covalently joined. Recall milk contains a trace of glucose but lots of lactose. Does bakers yeast ferment milk. If you do your work carefully using lots of yeast, not just one crumb, the yeast/milk should be safe to taste, provided a qualified cook (your mother?) helps you.

— If you have a interest in becoming a bio-scientist, you could memorize this info about starch, glucose, sucrose, lactose, maltose. This one is easy lactose = glucose + galactose. The ability to recongnize the importance of learning such things will help others know you have aptitude and interest in biochemistry.

Testing the ability of bakers yeast to grow on three carbohydrates.

— Dissolve 10 mL of cornsyrup per liter of tap water. While it is not pure, label it "Glucose Water."
— Dissolve 10 mL of table sugar per liter of tap water and label it "Sucrose Water."
— Dissolve 10 mL of starch per liter of tap water and label it "Starch Water."
— Dissolve 10 mL of dry milk per liter or use skim milk if you have it and label "Milk". In microbiology experiments we prefer skim milk to avoid the little globules (balls) of fat which make the solution turbid.
— Prepare your inoculum by dissolving yeast in tap water. Try a couple matchheads amount in 15 or 20 drops of water.
— This experiment would be more informative if you sterilize your medium.

Additional suggestions:

— you can use 500 ml or other size soda pop bottles to measure volumes.
— 20 drops of water equal about 1 mL
— begin collecting pop bottles and other useful things for your home microbiology lab.

Experiment 3-Using A Complete Medium

Corn syrup water is not a very good medium because it is low in nitrogen. All living things require amino acids to make their proteins. Some proteins make up muscle or other structures; other proteins are called enzymes. Enzymes are catalysts which reduce the energy level required for reactions. Thus, fire at high temperature burns sugar, but using enzymes some bacteria can burn (oxidize) sugar near the freezing point.

While all organisms require nitrogen, very few can use nitrogen directly from the air. Plants can use inorganic salts called nitrates and some can use ammonium salts. Many bacteria can use ammonium, nitrite, or nitrate salts and some can use nitrogren direct from the air. Many microbes and humans can use only amino acids. Humans can eat proteins and digest them down to amino acids. It is hard to find sources of free amino acids around the home, but there are lots of sources of proteins including egg white, milk, meat, and beans (has other things also).

Since some microbes require amino acids, common microbial media contain proteins, peptides, or amino acids. Peptides are partially digested proteins. We could add milk to our corn syrup medium. Milk alone makes a good medium for some bacteria because it contains proteins and lactose, a sugar. The white color of milk makes it difficult to see the growth of the bacteria unless the bacterium has a strong color. Meat broth is a good way to add proteins, peptides, and amino acids to a medium.

Using these ideas you can try to make a more complete medium for bacteria. For now you can try some experiments on your own, such as adding soil to milk and incubating at 4C, cold room, normal room, or higher temperatures. Expect some very strong odors. If you screw the lids on tight, there is a risk that gas pressures could build so high that the bottle or tube will burst. Don't overlook the use of plastic bags for cultures, but they may leak.

Experiment 4-Selective Medium

Many kinds of food are made using selective medium and bacterial fermentations. Salt or vinegar is often added to prevent the growth of most kinds of bacteria. Try adding varying amounts of salt to tubes of a good medium for yeast add gas collection tube if you have them, does the salt interfere with growth of the yeast. Some yeasts grow at remarkably high salt levels.

Try adding vinegar in varying amounts to your proven yeast medium. In general, yeasts are noted for growing at acidic pH. If you don't have a pH indicator, you might try red cabbage juice. Here is a cabbage experiment. Slice some cabbage finely. Pack it tightly into 3 small jars. Add dilute vinegar to jar-1. Add dilute salt water to Jar-2. Add nothing to jar-3. Put the lids on loosely and incubate at room temperature or a little cooler. Jar-3 will rot quite quickly. Jars-1 and -2 will rot but less so.

For an interesting project, get a canning book and the help of your mother, grandmother, or any experienced canner and try making fermented pickles or sauerkraut using such selective methods. The food is safe to eat only if you follow the directions carefully—it is safer to have an expert cook with canning experience to help you.

Most advanced microbial experiments require the use of solid media for isolation of pure cultures and the use of sterile medium. Pasteur accomplished much using selective methods, but he and others began using sterile medium and then rapid advances began.

— Mix some corn syrup, or table sugar, or meat broth with tap water (or water from a clear puddle). Use all three if you wish. This will be your medium for the growth of your microbe. It is best if your medium is very clear (free of particles). It is better if it is close to water white so don't use black corn syrup unless that is all you have. Do not use too much sugar because too much sugar will interfere with the growth of your microbe.
— The medium will contain lots of bacteria and mold spores. Boiling for a while will kill most of the growing bacteria (vegetative cells) but not all of the spores.
— in each container.

— Put some food in each container
— Observe examples of microbial growth around you.
— Selective growth of an organism. vingegar, salt.
— Notice that when microbes multiply gets cludy, corn syrup in water and inoculate.
— sterile toothpicks tostart rotten apples
— tyddalize milk 3 days in a row save a conrtol

SAFETY RULES FOR MICROBIOLOGY

Be sure to keep your experiments away from small childern. Work with your parents to plan a place to keep your things so that small kids or unsuspecting adults do not drink or eat your experiments because they thought it was food. This means you should not keep your experiments in the refrigerator or on food shelves. You can keep them in suitable boxes with a sign so they are will marked.

Do not sniff moldy items. There is a chance that the spores could get into your longs and begin to grow. Certain fungi can grow in our lungs and antibiotics do not help much. Death has a good chance of occuring if you get infected by such fungi. Actually, one cubic centimeter of air in homes commonly has 5,000 or 8,000 particles of dust and some of these are mold spores including the dangerous ones. Each good deep breath contains about 20 million particles.

Don't waste too much time wishing for equipment you don't have. Spend your time thinking about things you can do with what you have. The begin some simple experiments and you will learn what you need. Later you may want to buy some supplies, but at first see how much you can do with items from your home, farm, and gifts of discarded items.

— B021 - First do some simple experiments to learn what you need. Working with non sterile equipment.ooker, loop, sterile sticks or paper strips.

— B022 - Making a meat broth medium.
— B023 - Tubing your media
— B024 - Sterlizing and storing your media

These experiments still need lots of rewriting, but the facts are correct. If you were able to sterilize your medium. Save a culture of the yeast for your collection, but it is just as well to wait until we can isolate a culture by streaking in a later experiment.

If you have a pressure cooker, you can use it to kill all the microbes, including spores of all kinds and viruses.

If you have a pressure cooker, put your liquid medium in a glass jar with the lid on loosely and cook (autoclave) at 15 pounds pressure for 15 minutes. Read the manual for the pressure cooker to learn the setting for 15 pounds and read the safety precautions. Get the owner of the cooker to supervise your work. It is possible for a pressure cooker lid to fly off and injure you. Do not forget to p

In fact, the early humans and all animals and plants are involved with microbes. With each breath, you inhale about 5 million to 20 million dust particles and many of those are microbes.

If Your Yeast did not Grow on Starch

— recall that your saliva contains an enzyme called salivary amylase which digests starch. does it digest starch enough to allow yeast to grow on starch.
— recall that acids catalyze hydolysis of starches, proteins and other condensation polymers. Add a drop of acid. battery acid, muractic acid, vinegar. and place the tube in boiling water for a while.
— can you think of other ways to dryrlize starch, such as barley sprouts (malt)

General behaviour

No running or "horse play". Try to anticipate potential problems.

Food and drink consumption

Food/drink cannot be stored (even in lockers or in bags) in the laboratories or areas where laboratory-related work is carried out or where organisms or chemicals are stored. This includes cold-rooms and refrigerators in laboratory areas.

Food/drink may be stored and consumed in offices that open directly into laboratories if:

a) the laboratory is not an OGTR certified PC2 lab,
b) the laboratory is not PC3,
c) the laboratory is not a QAP, and
d) a valid risk assessment has been completed to show that there is an acceptable risk associated with this.

Food/drink (even empty cups and containers) are not permitted in OGTR registered PC2 laboratories or in offices opening into them (OGTR regulations). A valid risk assessment puts the onus onto the supervisor to ensure that there is an acceptable level of risk, and that suitable precautions are formulated, documented, and adhered to.

Smoking, applying cosmetics, etc

No smoking is permitted anywhere in the building and in some designated (with signs) areas close to the building. Chewing gum, sucking confectionary, and applying cosmetics are governed by the same rules as eating and food storage.

Laboratory gowns or labcoats

Laboratory gowns or coats must be of sufficient length to protect the worker (i.e., down to about the wearer's knees), have long sleeves, and elastic or restricted cuffs. They must close at the back or, if front-closing, have enough overlap at the front to prevent gaping.

Approved laboratory gowns must be worn in all laboratories and laboratory-like areas at all times. This is an absolute requirement for work in OGTR labs (OGTR regulations). It is a requirement for all people working in any laboratory or laboratory-like area in this department to be wearing an approved gown, except at approved writing areas. Laboratory gowns must be removed before entering "clean areas", eg., the tea room, stores, media, toilets, library, office areas, etc.

We recognise that some people have writing areas in laboratory spaces. In most cases it would be more hazardous to wear a laboratory gown at these after working at the bench than not to wear a laboratory gown. Each desk should be assessed individually, so that specific hazards are identified. A risk assessment and the segregation of activities SOP will document that this has been done.

Suitable precautions should be formulated, documented, and adhered to. The risk assessment would include advice like: the need for hand washing, no gowns/ gloves, clear separation of activities, etc.

Safety glasses

Safety glasses must be worn for all activities in the laboratory, unless a risk assessment for that specific activity shows that the risks associated with not wearing them are acceptable. Safety glasses and face shields must conform to AS 1337. Prescription glasses that are not too small may suffice in most cases, but a pair of safety glasses should be worn in addition when carrying out hazardous activities. Prescription glasses do not offer side protection and may not be strong enough, and so do not offer the same degree of protection as safety glasses do. Remember, an accident may occur in another part of the room, and still affect you. Full face shields must be worn for all work involving corrosive or toxic liquids, ultra-violet light, liquid nitrogen, or when disposing of broken glass into the waste bins outside the loading bays.

A generic assessment cannot be made of the risk of not wearing safety glasses; instead, each hazard has to be assessed in the context of not wearing safety glasses. So, each activity has to be assessed with the appropriate PPE and precautions, including the hazard of injury to a person in the room who is not wearing safety glasses. It is the responsibility of the supervisor who approves the protocols that are developed from the risk assessments. It is probably easier to wear safety glasses at all times while working in the laboratory.

Some people have writing areas in laboratory spaces. It is uncomfortable to wear safety glasses while writing or working at a computer, and the practice could lead to vision problems. The best way out of this is to assess the work at the lab desk as part of the assessment for lab gowns. Each desk should be assessed individually, so that specific hazards that could lead to eye injuries are identified (for example, nearby equipment and procedures). A risk assessment will document that this has been done. It may be easier to change the location of equipment and procedures, than to make someone wear safety glasses all the time.

Disposable Laboratory Gloves

Disposable laboratory gloves are not to be worn in communal areas. Door handles, telephones, computer keyboards and mice (except in clearly labelled circumstances), lift buttons, etc. are not to be touched with gloves. If needed, wear one glove and use the un-gloved hand to open doors, operate lifts etc. Rubber or disposable gloves should be worn when handling/working with:

— human blood or other body fluids
— dangerous chemicals

— infectious, or potentially infectious materials
— U/V light boxes
— radio isotopes

Hand washing

Hands should be washed after completing each task and always before leaving the laboratory.

Clothing, footwear, hair and electronic devices

Clothing and footwear must be suitable for laboratory conditions, i.e., clothing must not present any hazards by being too loose, accessories must not present a hazard during laboratory work, etc., and footwear must be fully enclosed. If you want to wear other shoes to and from work, etc., you might want to keep an old pair of fully enclosed shoes at work to wear in the lab. Thongs, or bare-backed sandals or shoes, or very high heels must not be worn in the laboratory. Bare feet are prohibited in the building. Long hair must be tied back to avoid contact with microorganisms and equipment.

Mobile phones, iPods, radios and similar electronic devices should be used judiciously. These may become contaminated and should be restricted to writing areas, and handled after PPE has been removed and hands washed. They may also distract workers and therefore cause safety problems. The laboratory head or area supervisor may prohibit the use of these items in situations where it may be considered unsafe, lead to contamination, compromise quarantine or other regulations, or affect productivity or harmony in the workplace.

Avoiding ingesting microorganisms

The best protection to ingesting microorganisms is not to put them in your mouth. Labels and envelopes must not be licked. Pencils and pens must not be placed in the mouth.

Chewing of fingernails, playing with hair, applying lipstick, eating, drinking, etc., are not allowed. Wash your hands when leaving the laboratory to go to lunch, etc.

Pipetting

Do not pipette by mouth. The use of pipettes with cotton plugs to reduce contamination is preferable. Place pipettes in disinfectant solution tip-first to minimise aerosol production. Submerge for 18-24 hours. Residual volumes from pipettes create aerosols; use mechanical devices that are calibrated to deliver. Replace glass pasteur pipettes with soft plastic varieties

Syringes and needles

Restrict the use of "sharps" to a minimum. Do not bend needles, or try to replace the caps after use. Use syringes fitted with blunt cannulas where possible. Avoid using syringes to mix infectious liquids (if essential, hold the tip of the needle under the surface of the fluid and avoid excessive force). Discard used syringes and needles into an approved sharps container.

Use of Aspirators

Vacuum aspirators are commonly in use in biosafety cabinets for removal of supernatants and filtering liquids, etc that may contain infectious agents. The following guidelines should be followed when setting up and operating aspirators:

— the vacuum outlet must have a HEPA filter to prevent infectious aerosols being exhausted

— the vacuum pump should be turned on prior to attaching the tubing to ensure that any liquid collecting in the tube between uses is removed. This reduces the risk of droplets forming and creating a splashing hazard while handling or attaching tubing.

— upon completion of each use the pump should be run for a sufficient time to allow liquid to be collected into the trap. The tubing should be fastened in an elevated position to ensure any residual liquid drains down the tube rather than collecting as droplets at the end of the tubing causing a contamination risk.

— the tubing should not be removed from the biosafety cabinet between uses. At the completion of work the tubing should be decontaminated with 80% v:v ethanol run through the lines.

— vacuum traps must only be opened and replaced within the biosafety cabinet. Aerosols and droplets are readily generated when changing the lids on the traps. Liquid waste must be decontaminated by autoclaving or appropriate chemical treatment prior to disposal.

Chemical disinfectants must not be autoclaved. Chemically treated waste must be tipped down the sink. Chlorine based disinfectants (bleach and diversol) are strongly corrosive and over time will corrode some types of vacuum pump severely reducing their working life if added to vacuum traps prior to collecting waste.

Hazardous Chemical/biological spills

Hazardous chemical and biological spills and blood spills on floors, benches or equipment should be cleaned up immediately. Special treatment is required for spills of a biohazardous nature. Decontamination kits are available on each floor in the corridor.

Faulty equipment

Any faulty equipment should be removed from service for repair or disposal.

Corridors and exits

Corridors and exits must not be obstructed by equipment or stored items.

Reporting accidents and incidents

All accidents and incidents must be reported on an Incident Report Form and forwarded to the Department Manager immediately.

Bench work

Minimise aerosol production when carrying out work on an open bench. Use equipment designed to contain aerosols for all manipulations such as shaking, mixing or sonication. A period of five minutes is required after these manipulations to allow aerosols to settle. If appropriate, subculture organisms in biological safety cabinets. Place "bench coat" or a similar absorbent paper over working surfaces and autoclave after use. Replace this regularly. Decontaminate work benches after spills and after work has been completed.

Bacteriological loops

When flaming wire loops, draw the loop gradually from the cooler to the hotter part of the flame to minimise spattering, or use electric heaters. Ensure the loop is completely closed and the loop wire is not longer than 6 cm. Disposable plastic loops must be placed loop-end down in disinfectant for 18-24 hours.

Fungal spores

Petri dish cultures of fungi should be sealed and incubated with the lid uppermost to prevent the dispersal of fungal spores. Recognise fungi as potential pathogens and be aware of the ability of some species to produce mycotoxins.

Petri dishes

Take care when handling Petri dishes that contain condensate. This may contain viable microorganisms that can be spread via droplets or aerosols when the plates are opened or dropped.

Tissue grinders and homogenisers

Open and operate tissue grinders in a biological safety cabinet. Hold glass grinders in a wad of absorbent material and wear gloves. Wait 10 minutes before opening a blender bowl to allow aerosols to settle. Refrigerate to condense aerosols. Use models designed to prevent leakage from rotor bearings and O-ring gaskets or use a "stomacher".

Workstations/computers in laboratory areas

If possible, do not place computers in laboratories. If this is required to run equipment or log data, or is otherwise unavoidable, reorganise the laboratory so that equipment that may harm people working on computers, or damage the computers themselves, is not placed near the computers. Install perspex screens if required. Sealed and/or quiet equipment, preferably without water or chemical use, is generally better than equipment that produces (or potentially produces) steam, vapours, aerosols, chemical spills, biological spills, radioactive spills, heat, dust or noise.

LABORATORY TESTS

Quality Assessment

Quality assurance, is defined by International Organization for Standardization (ISO), is all those planned and systematic actions necessary to provide confidence that a product or service will satisfy given requirements for quality. Such a system comprises two major parts: Quality Control which covers the tangible procedures performed in the laboratory on a regular basis to ensure that all aspects of the daily work are under control and Quality Assessment. Quality control will include monitoring of procedures, consumables, equipment, temperatures of incubators, refrigerators, and staff performance.

In order that a laboratory can assess its performance the second part must be implemented: participation in a Quality Assessment or Proficiency Testing scheme whereby the Quality Control procedures are challenged by the examination of samples of known but undisclosed content. The operational techniques and activities that are used to fulfil requirements for quality are also referred to as Analytical Quality Control (AQC) and can be differentiated into three lines of checking as outlined in Table 1.

Table 1 Analytical control in microbiology

Line of checking	Responsibility	Frequency	Purpose
First	Analyst	High	All aspects of analysts under control and consistent over time.
Second	Person independent of the analyst	Less frequent	Different analysts or equipment produce similar results. Individual results not biased.
Third	Laboratory management	Regular intervals	To nsure inter-laboratory standardisation

The first line of checking is a means of self control by the analyst, but it should be supervised by the direct superior responsible for setting criteria and defining action plans. It should be included with every series of analysis. First line checks can be divided into those to be undertaken a) before the analysis (samples, equipment, media, filters, and reagents), b) during the analysis (noting of all information that becomes available such as temperature, anaerobic conditions, confirmation rates, colonial appearance, background flora, etc.) and, c) in addition to the analysis. The latter would include internal quality control procedures such as examination of additional samples with known characteristics, parallel plating, procedural blanks, positive and negative control samples, colony counts on different volumes/dilutions, use of control charts, and use of sufficient colonies for confirmatory tests.

The second line checks are implemented to assure reproducibility between different analysts or equipment, during training of new workers, and evaluation of established staff in order to maintain standards of subjective interpretation. Such second line checks would include, a) duplicate counting (randomly selected samples) by the same person to provide the counting error under repeatability conditions and by different persons, thus including both random and systematic components to the variation. These will allow relative differences or standard deviations to be computed, b) duplicate analytical procedures to test the whole quantitative procedure by using duplicate samples and then plotting control charts, and c) intensified quality control tests as listed for first line checking.

The third line checks should be supervised by the quality assurance officer and include participation in an EQA scheme (proficiency testing) and the use of certified reference materials (CRMs). In the former, the samples are examined by different laboratories, the results interpreted retrospectively by the central organisation, and the performance compared with other participants. It is a flexible approach whereby participants apply their own laboratory methods. With CRMs, all laboratories follow a strict protocol and the certified value is valid only for the applied method. Results obtained with other methods can be compared with the certified values.

EXTERNAL QUALITY ASSESSMENT (EQA) OR PROFICIENCY TESTING (PT)

The requirement for such schemes and the types of scheme available will be discussed with specific emphasis on those provided by the Public Health Laboratory Service (PHLS), UK, in the area of food microbiology. EQA can be defined as an independent assessment of the competence of laboratories to perform tests accurately and precisely, providing a challenge to the effectiveness of the quality system of a laboratory and adherence by staff to that system.

The PHLS organises microbiology EQA schemes for food, water, and clinical laboratories. The latter scheme is well established and has been operating for more than 25 years. The food and water schemes were launched in the early 1990s. The PHLS food scheme was in part developed as a response to the requirements of food legislation and the need for mutual recognition of results between member states of the European Single Market.

For the concept of mutual recognition of results there is a need for mutual recognition of microbiological standards, guidelines and specifications, the use of standard methods of known accuracy and precision, laboratories competent to perform the agreed tests accurately and precisely, and periodical, independent assessment of accuracy and precision of microbiological results (proficiency testing). It is the latter requirement to which EQA schemes are targeted. There are many benefits to be derived from participation in an EQA scheme beyond the requirement for official laboratories stipulated in the EU legislation. Such benefits are listed below:

— With Quality Control provides a total Quality Assurance package;

- Assessment of level of performance against an external standard of performance;
- Comparison of current and past performance;
- Comparison of performance with other laboratories;
- Identification of unsatisfactory performance (laboratory personnel);
- Method of demonstrating staff competence;
- Assessment of improvement over time;
- Identification of declining performance and introduction of remedial action;
- Educational tool for training staff and improving performance;
- Performance of less familiar tests can be improved;
- New methods can be introduced;
- Source of help and advice;
- Strict confidentiality allows discussion of problems;
- Provides confidence to participants;
- May assist in the marketing of laboratory services.

PCR TESTING AND ANALYSIS

For the last 150 years, fungi have been identified based on macro and microscopic examination of morphological characteristics and biochemical assays. This would often entail growing the fungi on synthetic media for long periods of time at varying temperatures to achieve sporulation and the production of conidia. Often this could take upwards of several weeks to a month to complete.

Recently, with the development of more rapid molecular based techniques to identify microorganisms appearing in journal after journal, researchers at the US EPA's National Exposure Research Center have developed and patented a method for the rapid identification of common indoor air fungal contaminants.

They identified unique Deoxyribonucleic acid (DNA) sequences for many of the problematic fungal species and designed a molecular based assay to identify and quantify these organisms rapidly. This licensed technology will rapidly replace the subjective identification of fungi by optical microscopy. Since this technique is based on the identification of unique DNA sequences and not the experience of the microscopist conducting the analysis, it is accurate, reproducible, and very rapid.

The multi-step PCR testing and analysis process begins with the sampling, which can be achieved utilizing standard field sampling techniques (i.e. swab, tape-lift, Air-O-Cell, 37mm PCM cassette filter).

Next, the sample is processed and DNA is extracted from the samples. This extracted DNA is then added to a PCR cocktail or "master mix". This master mix includes the primers specifically designed for the fungal species to be identified and enumerated.

It is then placed into a thermocycler for amplification. The first stage of the PCR reaction is the denaturation or melting of the DNA (the double helix splits apart). This is achieved by heating the sample to 95 degrees Celsius. Next, the mixture is cooled to 45-60 degrees Celsius. This is specific to the primer sets and the annealing stage occurs. The two single stranded primers find their complementary DNA sequences and hybridize or bond to them.

The third stage of PCR amplification is extension or synthesis. The thermostable polymerase (Taq) catalyzes the addition of complementary bases to the single stranded template DNA. This forms two new double-stranded DNA helix. These three stages of PCR amplification are then repeated over and over (cycles up to 40 times) again. Hence, the Chain reaction.

PCR testing has become commonplace in research facilities across the world. In the 1990's the commercial release of quantitative PCR (Q-PCR) marked the beginning of a new era in rapid molecular diagnostics. Q-PCR, commonly referred to as Real time PCR, has come to the forefront of the diagnostic and research world by providing accurate quantification of DNA in real time, while decreasing the time necessary to produce detectable levels of PCR product. With this, researchers began to apply this technology to rapid molecular level microorganism and genetic modification detection systems.

HEPATITIS A TESTING

Hepatitis A is caused by the hepatitis A virus (HAV), which is spread through infected food that has been contaminated with fecal material. HAV transmission by food commonly occurs through some local source of contaminated water, surfaces, and infected food handlers. Hepatitis A outbreaks have also been associated with consumption of fresh produce contaminated with HAV during cultivation, harvesting, processing, or distribution.

Cooking minimizes the risk of illness by reducing or eliminating the virus or other pathogens. Contaminated foods that do not undergo cooking, or those that are lightly cooked, such as fruits and vegetables, are often at greater risk of transmitting Hepatitis A. HAV is a virus that resists drying and is more heat resistant than many other enteric viruses. Properly cooked foods are of less risk in the transmission of Hepatitis A.

Recently developed techniques using polymerase chain reaction (PCR) assays to amplify the nucleic acids of a pathogen have been modified for detection of HAV in produce. HAV is a positive strand RNA virus belonging to the hepatovirus group of the picornavirus family. Using total RNA extracted from captured HAV particles and reverse transcription-PCR, this procedure can be applied for routine surveillance of HAV in fresh produce.

During this stage of analysis, scientists will perform one of two testing techniques—Fourier Transform Infrared (FTIR) Spectroscopy and Scanning Electron Microscopy (SEM) with Energy Dispersive Spectroscopy (EDS), one that is best for organic materials, and one that is best for inorganic materials.

CONVENTIONAL LABORATORY METHODS FOR CYANOTOXINS

Over recent years it has become apparent that toxic cyanobacterial blooms are on the increase, presenting a hazard to animal and human health. The importance of algal toxins is reflected in their inclusion of EPA recognised contaminants in water. Microcystins have been extensively studied and reported over recent years. Despite the number of microcystin variants and lack of standards, a large number of biological and chemical methods have been optimised for a variety of matrices, usually cells, water and tissue. Methods developed for microcystins are suitable for the pentapeptide nodularins, although these cyanotoxins usually occur in brackish water.

In contrast, relatively little work has been done on methods detection of other known toxins, anatoxins, cylindrospermopsins, BMAA and aplysiatoxins. Saxitoxins being the exception, as they occur widely in the marine environment and many methods have been developed for their detection in shellfish. However, there has been only limited application of these methods to freshwater samples. There are many challenges in assessing and selecting suitable methods since blooms can not only be composed of co-occurring species but it is also known that some species produce multiple classes of toxins.

ANALYSIS OF MICROCYSTINS

Microcystins are the most commonly reported cyanobacterial toxins. Although nodularins are less of a problem in freshwater, most methods developed for microcystins are suitable for nodularins. By far the greatest challenge in analysing microcystins is the fact that there are in excess of 65 variants characterised to date and most likely others yet to be identified. It is essential that any method used has the ability to detect all variants, regardless of availability of standards. Equally important, extraction and separation procedures must be suitable for the chemical range of variants in order to obtain accurate qualitative and quantitative data.

HPLC METHODS

There are many liquid chromatography based methods in the literature, utilising a range of stationary phases, mobile phases and detectors for both isocratic and gradient separations. However, reversed-phase chromatography with diode array detection (HPLC-PDA) has been the most widely used approach over the last two decades, as it enables detection of all microcystins based on their characteristic UV spectra. Use of a gradient helps to ensures microcystins variants will be separated and despite lack of standards or certified reference materials, quantification of approximate total microcystin content is possible based on purified MC-LR to give MC-LR equivalence.

Inter-laboratory validation data supports this approach combined with concentration and clean-up on SPE (Isolute C18) for the monitoring of intra and extra cellular microcystins in water samples as recommended in a "Blue Book" publication in the UK. Limits of quantification reported are 1-10 ng on column. A recent inter-laboratory trial highlighted the need for certified reference materials as commercial material that is currently available is essentially a laboratory reagent not a standard. When this material is used as a standard it results in varying responses for the same samples in different laborato-ries. Detection limits have been improved by the use of immunoaffinity SPE for concentration, however, there are still limitations on binding capacity and the volumes loaded which must be overcome if this is to be a practical solution.

Recent advances, using recombinant antibody fragments, have demonstrated potential for the development of cost effective, robust and reproducible immunoaffinity cartridges. As technology has evolved, LC-ESI-MS or LC-ESI-MS/MS is becoming the preferred technique as it offers greater selectivity and sensitivity than diode array detection. Good sensitivity was achieved using a single quadrupole (LC-ESI-MS), LOD of 11, 72, 21 and 6 pg for MC-LR, MC-RR, MC-YR and nodularin respectively on column (1 mm I.D.) using selected ion monitoring (SIM). However, most methods published in the literature use tandem MS, which enables noise reduction and thus greater sensitivity, multiple reaction monitoring, and the removal of the need for complete separation of analytes.

This approach enabled the development of a high through-put method which analysed ten microcystins in 2.8 minutes, without the need for complete resolution. However, although the potential of LC-MS/MS is unequivocal, much work is still needed since most methods have been developed with a limited number of microcystins and there is no way to guarantee detection of unknown microcystins as fragmentation patterns vary considerably with conditions and microcystin chemistry itself.

Maximizing individual microcystin sensitivity can be achieved by complex methods utilising time scheduled selected reaction monitoring conditions as demonstrated by Bogialli et al. Several reports have examined matrix effects on analysis of tissue samples, illustrating the importance of the inclusion of this work for any method under development and in subsequent validation. Ruiz et al. demonstrated a 15% over estimation of MC-RR in extract from kidney compared to a 37% decrease in detection in liver. From these findings they recommended the use of matrix matched standards for use when quantifying unknown samples.

Matrix-assisted laser desorption (MALDI) has been used in conjunction with TOF analysers for the detection of microcystins and unknown variants in small samples (Welker et al. 2002). Characteristic fragmentation was achieved by post-source decay, which results in destruction of the peptide bonds. Whilst rapid, this offline technique requires some extraction to eliminate matrix/sample interferences, but, as improved matrices are developed, there is future potential for an approach eliminating time consuming sample preparation and chromatography. This is illustrated in a recent publication, describing the use of MALDI linked to a triple quadruple for the qualitative and quantitative determination of spirolide toxins. The combination of this ionisation technique with sensitive multiple reaction monitoring, proved to be precise and accurate without the need for extensive sample preparation.

Another exciting approach which is rapid and eliminates time consuming SPE, where the microcystins were captured on a hydrophobic chip and subsequently ionised by surface-enhanced laser desorption ionisation-time-of flight MS (SELDI-TOF-MS) enabled determination of 2.5 pg MC-LR in 2 μl (1.2 μg L^{-1}) water. However, severe matrix effects were experienced when more complex samples were analysed, and it was not possible to monitor the characteristic *m/z* 135 due to background interference. Future chip developments could present the way forward although may prove costly.

IN-VITRO BIOASSAYS

To compliment the large number of physico-chemical methods there exists a significant number of bioassays for detection of microcystins. Microcystins and nodularins are strong inhibitors of protein phosphatases, PP-1, PP-2A and

PP-3, PP-2A being the most sensitive. This functionality has been exploited to develop assays which provide a direct measure of toxicity. A range of substrates have been used but the most commonly used are p-nitrophenol phosphate (P-NPP), 4-methyl-umbelliferyl phosphate (MUP) and 6,8-difluoro-4-methylumbelliferyl phosphate (DiFMUP).

The latter has been successfully validated against HPLC and mouse bioassay for the detection of okadaic acid in shellfish. This approach has been adapted to a rapid microplate assay for screening microcystins in drinking water without the need for pre-concentration, achieving a detection limit of 0.1 μg L^{-1}, which is well below the provisional guideline value. This assay provides a useful pre- or post analytical screen for bioactivity although false positives may be obtained from other phosphatase inhibitors, which may occur in environmental samples. Many researchers have reported good correlation of data obtained by protein phosphatase inhibition assay and HPLC-PDA. The necessary components are available commercially although there is batch variation in enzyme activity.

IMMUNOASSAYS

Immunoassays, exploiting polyclonal, monoclonal antibodies and recombinant antibody fragments, are widely used as screening tools for microcystins and nodularins and are well reviewed elsewhere. Several kits are commercially available, in microtitre plate or tube format. Many of the assays/kits use antibodies raised against MC-LR and subsequently may have limited cross reactivity (EnviroLogix Inc, Portland, ME, USA), whereas kits using antibodies raised against ADDA provide improved sensitivity and excellent cross-reactivity. However, the behaviour of non-toxic degradation products including free ADDA is as yet unknown.

These ELISA kits are supplied in a 96-well microplate format with ready to use reagents enabling screening of up to 96 samples in 2.5 hours with a consumable cost of $400.00. All of these commercial kits are simple to use, rapid and economical for screening.

OTHER USEFUL METHODS

A cost effective, rapid, thin layer chromatography (TLC) method has also been developed which enables detection of microcystins to meet the WHO 1 µg L^{-1} guideline.

This method relied on visualisation of the microcystins on the developed TLC plate using N,N-dimethyl-1,4-phenylendia-monium dichloride (N,N,-DPDD) and good correlation was achieved compared to protein phosphatase and ELISA assay. However, without sophisticated spotting and scanning devices, this is not quantitative, but would serve as a useful screen for known microcystins, although it does require improvements in sample concentration to remove interfering contaminants.

A method for determination of total microcystins relies on oxidation to produce 2-methyl-3-methoxy-4-phenyl-butyric acid (MMPB) from ADDA, which is detected by GC-MS, HPLC-Fl or HPLC-TSP. Whilst this method has been demonstrated to be useful for complex samples such as sediments, the need for oxidation, and the fact that only total microcystin is determined, make it a complex, time consuming and expensive screen.

Despite these disadvantages, this could be a useful confirmatory method and can be used with a wide range of instrumentation without the need for microcystin standards. Most methods described determine free microcystins, this method will also detect bound microcystin, thus providing a complete picture in metabolism studies.

Capillary electrophoresis based methods exploit high efficiency columns to separate variants often problematic in

LC separations such as MC-LR and MC-LR providing a useful complimentary technique. Issues such as sensitivity and interfering compounds have been overcome by improved online and offline sample clean-up.

COMBINED METHODS

HPLC-UV/PDA has been shown to be a powerful tool in combination with protein phosphatase inhibition or ELISA assay. HPLC-PP2A was first reported in 1991 as a highly sensitive bioscreen for okadaic acid along with related polyether toxins and later applied for the detection of microcystins in freshwater environments. These approaches are still used, often along side mass spectrometry to determine complete structure/activity profiles of unknown samples.

Fractionation into 96 well plates was used to increase automation and extending the assay to include an immunoassay providing LC-UV/ELISA /PP2A data, achieving detection limits 1000 x more sensitive than UV. Several groups have reported the use of ELISA alongside PP2A inhibition, providing a measure of total microcystins and toxicity, however, the most elegant use of these techniques is the immunophosphatase assay.

ANALYSIS OF SAXITOXINS

Saxitoxins (also known as paralytic shellfish poisons, PSPs) are another complex group of compounds which have presented a challenge over the last two decades. Until June 2005, the only validated method available was the mouse bioassay, routinely used for screening shellfish and phytoplankton.

HPLC ANALYSIS

In June 2005 an HPLC method relying on fluorescence detection of the oxidised saxitoxins was approved by AOAC after inter-laboratory validation. Whilst this method is robust, the sample processing is complex and two pre-column oxidation reactions/separations may be needed for quantification of the complete range of saxitoxins. A further problem is that oxidation of some GTXs, dcGTXs, dcSTX and dcNEO results in the production of two fluorescent compounds, thus requiring a broad range of standards. Despite the reported robustness, this is a time consuming and therefore expensive method. Automation of the derivatization procedure would reduce manual processing, however it must be noted the fluorescent products are not stable after a few hours.

An alternative approach using post-column derivatization has been preferred in many labs as it benefits from simple automation. However, three, more recently two, separations are needed to accurately quantify all toxins. This method is sensitive to changes in flow rate, reagent age and temperature. With both pre- and post column derivatization methods, it is ideal to run a sample without oxidisation to confirm peaks are not interfering contaminants.

Several methods using capillary electrophoresis have been reported although, they are not widely used and suffer from low sensitivity due to the low volume injected and the requirement for a very clean sample in order to obtain reproducible chromatography. It is reported that LOD is an order of magnitude greater than HPLC-FL/MS. Many assays have been described which exploit the functionality of the saxitoxins, i.e. sodium channel blocking activity.

REFERENCES

Ajello, L., L.K. Georg, W. Kaplan, and L.Kaufman, "Laboratory Manual for Medical Mycology", *U.S. Publ. Health Serv. Publ.*, 1963.

Angelotti, R. and MJ Foter, "A Direct Surface Agar Plate Laboratory Method for Quantitatively Detecting Bacterial Contamination On Nonporous Surfaces", *Amer J Public Health.* Sept:170-174, 1957.

Beneke, E.S., "Medical Mycology Laboratory Manual", Burgess Publ. Co.,Minneapolis, 1958.

Booth, C., "Fusarium", *Laboratory Guide to the Identification of the Major Species,* 1977.

Burleson F.G., T.M. Chambers and D.L. Wiedbrauk, *Virology: a laboratory manual,* Academic Press San Diego, Sydney, 1992.

Burgess, L. W., C. M. Liddell and B. A. Summerell, *Laboratory manual for Fusarium research,* 2nd ed. University of Sydney, Sydney, 1988.

9

Microbial Biodegradation

Biodegradation is the process by which organic substances are broken down by the enzymes produced by living organisms. The term is often used in relation to ecology, waste management and environmental remediation (bioremediation). Organic material can be degraded aerobically, with oxygen, or anaerobically, without oxygen. A term related to biodegradation is biomineralisation, in which organic matter is converted into minerals.

Biodegradable matter is generally organic material such as plant and animal matter and other substances originating from living organisms, or artificial materials that are similar enough to plant and animal matter to be put to use by microorganisms. Some microorganisms have the astonishing, naturally occurring, microbial catabolic diversity to degrade, transform or accumulate a huge range of compounds including hydrocarbons (e.g. oil), polychlorinated biphenyls (PCBs), polyaromatic hydrocarbons (PAHs), pharmaceutical substances, radionuclides and metals.

Major methodological breakthroughs in microbial biodegradation have enabled detailed genomic, metagenomic, proteomic, bioinformatic and other high-throughput analyses of environmentally relevant microorganisms providing unprecedented insights into key biodegradative pathways and the ability of microorganisms to adapt to changing environmental conditions.

Interest in the microbial biodegradation of pollutants has intensified in recent years as mankind strives to find sustainable ways to cleanup contaminated environments. These bioremediation and biotransformation methods endeavour to harness the astonishing, naturally occurring, ability of microbial xenobiotic metabolism to degrade, transform or accumulate a huge range of compounds including hydrocarbons (e.g. oil), polychlorinated biphenyls (PCBs), polyaromatic hydrocarbons (PAHs), pharmaceutical substances, radionuclides and metals.

Major methodological breakthroughs in recent years have enabled detailed genomic, metagenomic, proteomic, bioinformatic and other high-throughput analyses of environmentally relevant microorganisms providing unprecedented insights into key biodegradative pathways and the ability of organisms to adapt to changing environmental conditions.

The elimination of a wide range of pollutants and wastes from the environment is an absolute requirement to promote a sustainable development of our society with low environmental impact. Biological processes play a major role in the removal of contaminants and they take advantage of the astonishing catabolic versatility of microorganisms to degrade/convert such compounds. New methodological breakthroughs in sequencing, genomics, proteomics, bioinformatics and imaging are producing vast amounts of information.

In the field of Environmental Microbiology, genome-based global studies open a new era providing unprecedented *in silico* views of metabolic and regulatory networks, as well as clues to the evolution of degradation pathways and to the molecular adaptation strategies to changing environmental conditions. Functional genomic and metagenomic approaches are increasing our understanding of the relative importance of different pathways and regulatory networks to carbon flux in particular environments and for particular compounds and they will certainly accelerate the development of bioremediation technologies and biotransformation processes.

ENVIRONMENTAL POLLUTANTS

One consequence of the appearance of new synthetic organic chemicals in the environment has been the remarkably rapid evolution of microorganisms with new catabolic activities. For example, the extensive use of the herbicide atrazine has provided the selective pressure for the evolution of a new catabolic pathway for the degradation of this chemical. It has been suggested that the atrazine degradative pathway has been assembled recently, over a period of decades. Due to their metabolic versatility and evolutionary potential, microorganisms are able to utilise an amazing spectrum of noxious compounds, including halogenated chemicals such as 1,2-dichloroethane and vinyl chloride, and pesticides such as γ-hexachlorocyclohexane (lindane) and 1,2-dibromoethane.

Despite the fact that a variety of microorganisms can utilise these chemicals as growth substrate, many halogenated aliphatic compounds persist in the environment, which emphasises the inadequacy of the current catabolic capacity to deal with all recalcitrant pollutants. Gene technology, combined with the extensive knowledge base of the microbiology, biochemistry, and genetics of catabolic pathways, has been proposed to enable the engineering of new or improved catabolic activities for such pollutants.

Over the past decades research on the degradation of short-chain haloaliphatics has been focused mainly on chloromethanes, chloroethanes, and chloroethenes, since these compounds are produced industrially in large amounts and represent a widely recognised class of priority pollutants. Biodegradation of halogenated propanes has drawn less attention, although halopropanes such as 1,2-dichloropropane, 1,2,3-trichloropropane (TCP), and 1,2-dibromo-3-chloropropane, also form an important group of toxic and recalcitrant pollutants, in part due to their chemical and biological stability.

Thus far, no pure cultures of bacteria that are capable of aerobic growth on these compounds have been isolated by classical adaptation and enrichment techniques. Nevertheless, it is possible to design pathways on paper which bacteria could use for growing on such chemicals. The engineering of microorganisms for the degradation of recalcitrant halogenated aliphatic compounds is therefore a challenging goal.

C3-halogenated Aliphatics

Table 1 summarises a number of relevant C3-haloaliphatics which are of environmental and industrial importance. In terms of industrial production epichlorohydrin is the most important member of the C3-haloaliphatics, with a global production of about 700,000 ton/year. Epichlorohydrin is a versatile building block for the synthesis of epoxy resins, rubbers, adhesives, agrochemicals, and pharmaceuticals. In particular, optically active compounds have broad potential for synthesis. Commercial epoxide production occurs mainly via the chlorohydrin process with chloropropanols as intermediates.

Aqueous effluent emerging from epoxide production processes for the manufacture of propylene oxide or epichlorohydrin, for example, can contain considerable amounts of undesirable chloroaliphatic compounds, such as 1,2-dichloropropane and 1,2,3-trichloropropane, respectively, of which the removal is very expensive. A number C3-haloaliphatics have been used as constituents of soil fumigant formulations. The nematocides Shell D-D and Telone II comprise a mixture of 1,3-dichloropropene and 1,2-dichloropropane and have been used worldwide in agriculture. In addition, TCP was often found as a significant contaminant of these fumigant formulations. The extensive application of these soil fumigants has led to pollution of soils and waters in the Netherlands and United States.

Table 1. Industrial and agricultural uses of various C3-haloaliphatic compounds

Compound	Uses
1-Chloropropane	Solvent
1-Bromopropane	Intermediate in chemical synthesis of pharmaceuticals
1,2-Dichloropropane	Nematicidic soil fumigant, solvent, insecticide for stored grain
1,3-Dichloropropane polymerisation catalyst in the chemicals	Alkylating agent, ring-forming agent, synthesis of organic
1,2,3-Trichloropropane	Solvent, extractive agent, chemical intermediate in the production of polysulfone liquid polymers and dichloropropene, cross-linking agent in the synthesis of polysulfides
1,2-Dibromo-3-chloropropane	Nematicidic soil fumigant of citrus crops
Epichlorohydrin	Intermediate in the production of epoxy resins, synthetic glycerin, elastomers
1,3-Dichloro-2-propanol	
2,3-Dichloro-1-propanol	
	Precursor optically active compounds,
1,3-Dichloropropene	Nematicidic soil fumigant

The highly persistent brominated organochlorine compound 1,2-dibromo-3-chloropropane was used as a nematocide in citrus crops in the United States. It is 8 to 16 times as effective as the D-D mixture, and has been produced for agricultural use since 1955. In 1979, the United States Environmental Protection Agency banned almost all agricultural uses of 1,2-dibromo-3-chloropropane, which was triggered by a report on sterility in workers who had handled 1,2-dibromo-3-chloropropane in a pesticide division of a chemical plant in California. However, 1,2-dibromo-3-chloropropane is still detected as a groundwater contaminant in the United States, illustrating the persistence of this chemical.

Because of the chemical and biological stability of halopropanes, and to a lesser extent also C3-epoxides and halopropanols, these compounds can be widely distributed in the environment via water and air. Although these chemicals are often not toxic at the concentrations found in water and soil, they tend to accumulate in the tissues of higher animals and plants. Due to bioaccumulation deleterious effects may occur on organisms at or near the top of the food chain. The toxicity of these chemicals has been documented from surveys of employees handling these compounds and by tests with different laboratory animals. As a result they are now being recognised as potentially mutagenic, carcinogenic, and/or toxic to the liver, kidney, and to the nervous and reproductive system.

Causes of Recalcitrance

Many halogenated aliphatics are slowly removed in the environment. Some obnoxious cases of persistent chemicals include 1,1-dichloroethane, 1,1,1-trichloroethane, the chlorinated ethylenes cis- and trans-1,2-dichloroethene, trichloroethene, and tetrachloroethene, 1,2-dichloropropane and 1,2,3-trichloropropane. From a thermodynamic point of view, aerobic degradation of halogenated aliphatics should yield sufficient energy to sustain bacterial growth. However, due to the presence of the carbon-halogen bond these compounds are often too far from the mainstream of catabolic pathways to be used as a substrate.

Catabolic enzymes and regulatory proteins of existing metabolic routes do not properly recognise these xenobiotics, yielding insufficient flux of energy and intermediates that can be used for biosynthetic processes. This is not surprising, since microorganisms have been exposed to these chemicals only since about 100 years, a mere instant in evolutionary terms.

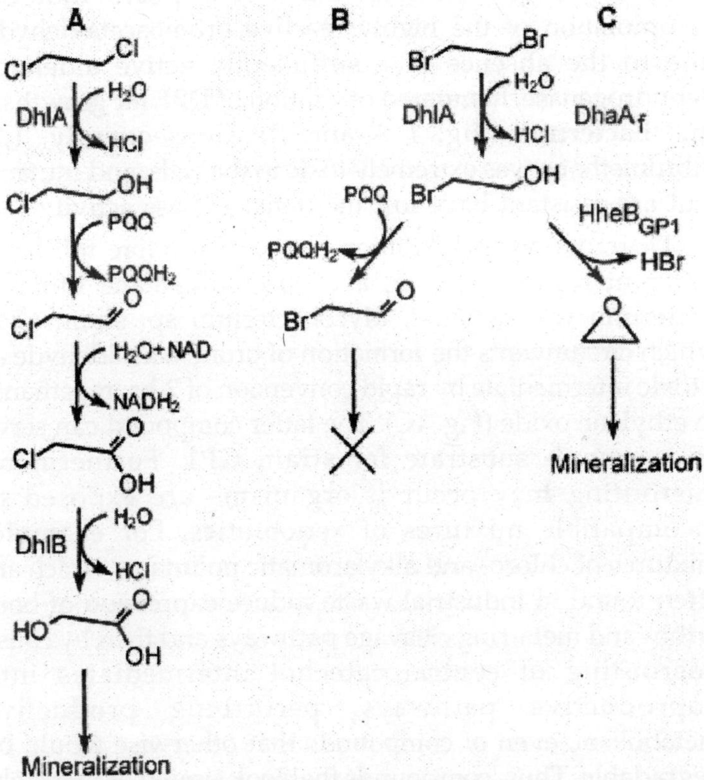

Figure 1. (A) Proposed route of the metabolism of 1,2-dichloroethane in X. autotrophicus GJ10. (B) Proposed degradation route of 1,2-dibromoethane in X. autotrophicus GJ10. (C) Proposed route of the metabolism of 1,2-dibromoethane in Mycobacterium sp. strain GP1.

For many xenobiotics, such as the highly toxic dioxins and PCBs, no effective degradative routes have been described. Often degradation of xenobiotics is incomplete or very slow, which may cause accumulation of reactive intermediates that destabilise microbial communities.

For example, the 1,2-dichloroethane (DCE) utilising bacterium X. autotrophicus GJ10 was not able to utilise 1,2-dibromoethane (DBE) as a growth substrate, although the organism seems to posses all the necessary enzymes for the complete degradation of DBE. It appears that the accumulation of the highly reactive. bromoacetaldehyde, due to the absence of a sufficiently active aldehyde dehydrogenase, hampered utilisation of DBE for growth by this bacterium (Fig. 1 A and B). Consequently, 1,2-dibromethane was extremely toxic to the cells, and mutants that are resistant have lost the dehalogenase activity.

Degradation of 1,2-dibromoethane therefore follows a different route. Recently, a 1,2-dibromoethane utilising bacterium was isolated, Mycobacterium sp. strain GP1, which circumvents the formation of bromoacetaldehyde as a toxic intermediate by rapid conversion of 2-bromoethanol to ethylene oxide (Fig. 1C). The latter compound can serve as a growth substrate for strain GP1. Furthermore, misrouting may occur if organisms are exposed to incompatible mixtures of xenobiotics. For example, mixtures of chloro- and alkylaromatic pollutants which are often found in industrial waste induce expression of both ortho- and meta-ring cleavage pathways and thereby cause misrouting of central catechol intermediates into unproductive pathways, perturbing productive metabolism, even of compounds that otherwise would be degradable. Thus, compounds that look similar at first sight may require very different routes.

A solely biochemical explanation is often not sufficient for understanding the observed persistence of halogenated compounds in the environment. The efficiency of degradation is also determined by various non-biological factors such as pH, temperature, salinity etc., since microorganisms usually operate within a fairly narrow range of physicochemical conditions.

Furthermore, microorganisms that in principle could derive energy from the degradation of halogenated compounds will not grow if the concentration of these compounds is below a certain threshold that is needed for the generation of new biomass. In that case, there will be no stimulation of growth of organisms that posses the required enzymes and the population of degraders will not increase. On the other hand, at high concentration direct toxic effects of an apolar chlorinated hydrocarbon can decrease the survival of degradative organisms. The limited bioavailability of apolar and poorly watersoluble compounds also can contribute to the persistence of xenobiotics in the environment.

Many compounds tend to adsorb onto soil particles or diffuse into clay structures and thereby become less available to microorganisms. Thus, understanding the bottlenecks of xenobiotic degradation requires a combined approach. The biochemical causes of the persistence of halogenated compounds may possibly be overcome by optimisation of enzymes and metabolic pathways in organisms relevant for biodegradation.

Natural Diversity for the Degradation of C3-Haloaliphatics

The use of microorganisms to degrade synthetic halogenated aliphatic compounds is an emerging technology. The success of this technology will in part depend on our ability to select or engineer microorganisms that can degrade compounds that are currently not biodegradable. For this, the naturally occurring microbial activities are the starting point.

Microbial Degradation of Halogenated Aliphatics

Aerobic microbial degradation processes can roughly be divided into two classes: cometabolic conversions and utilisation of the organic compound as a growth substrate. Mineralisation of halogenated aliphatic compounds, which always occurs if a compound support growth, is the most favorable way to eliminate environmental pollutants. An increasing number of aerobic microorganisms have been isolated that can utilise haloaliphatics as a growth substrate.

Some remarkable examples are bacteria that can grow on dichloromethane, 1,2-dichloroethane, and vinyl chloride, or on pesticides such as 1,2-dibromoethane, all of which are produced in large quantities by the chemical industry. Of the C3-halogenated aliphatics only a few mono- and dihalo-compounds are known to support microbial growth.

This includes organisms growing on epichlorohydrin, various halopropanols, and bacteria that degrade the soil fumigant 1,3-dichloropropene. Halogenated aliphatic hydrocarbons such as 1,2-dichloropropane (DCP), 1,2,3-trichloropropane (TCP), and 1,2-dibromo-3-chloropropane have not been shown to support growth under aerobic conditions.

Cometabolism of several short-chain halogenated aliphatics results from the lack of selectivity of some microbial enzymes. In this case, microorganisms convert a substrate without gaining energy or carbon building blocks for biosynthetic processes, and thus the organisms require the presence of a second substrate for growth.

For several environmental pollutants, such as chlorinated ethenes, cometabolism is at this moment the only available option for biodegradation under aerobic conditions. Cometabolic transformations are also important for the biodegradation of recalcitrant C3-haloaliphatic compounds. Some of the di- and trihalogenated compounds are thus far only known to be degraded by cometabolic transformations.

Enzymatic Dehalogenation of Halogenated Aliphatics

A critical step in a microbial pathway for the degradation of halogenated compounds is the cleavage of the carbon-halogen bond. This step may occur abiotically, but in most cases dehalogenation is enzymatically driven.

Haloalkane Dehalogenase

Haloalkane dehalogenases are involved in the initial dehalogenation of haloalkanes, and are thereby responsible for the capability of microorganisms to degrade these compounds. These enzymes catalyse the hydrolytic cleavage of carbon-halogen bonds in a variety of haloalkanes. Haloalkane dehalogenases have a globular structure and are composed of two domains: a large central catalytic domain with an α/β-hydrolase fold structure, and a second domain which lies like a cap on the main domain. The latter domain emerges as a large α-helical excursion between β-strands 6 and 7 of the catalytic core.

Different haloalkane dehalogenases show significant structural homology in the hydrolase core but diverge in the cap domain. The interface of the two domains defines the hydrophobic active site cavity containing the catalytic triad, which is composed of a nucleophilic aspartate, an invariant histidine and either an aspartate or a glutamate as the third member. Presently, the haloalkane dehalogenases of the strains Xanthobacter autotrophicus GJ10 (DhlA), Rhodococcus rhodochrous NCIMB 13064 (DhaA), and Sphinghomonas paucimobilis UT26 (LinB) have been extensively studied.

These enzymes display a broad substrate tolerance toward various chloro- and bromoalkanes, but are evolutionary optimised for the conversion of different haloalkanes. DhlA is most active on C2-C4 mono- and dihalogenated alkanes, whereas both DhaA and LinB recognise a range of C2-C9 cyclic-, mono-, di-, and polyhalogenated alkanes. At least 25 different halogenated aliphatics have been recognised as a substrate for one or more of these enzymes, including several environmentally important chemicals such as 1,3-dichloropropene, 1,2-dibromoethane, 1,2-dibromo-3-chloropropane, and 1,2-dichloropropane.

Haloalkane dehalogenase from R. rhodochrous NCIMB13064 (DhaA) is the first catabolic enzyme involved in the degradation of several C2-C8 n-haloalkanes. The dhaA gene encodes a protein of 293 amino acids which shares homology with DhlA in the hydrolase core, but the proteins diverge significantly in the cap domain. DhaA and DhlA have conserved amino acid residues involved in catalysis indicating that they have a similar reaction mechanism. Asp106, Glu130, and His272 form the DhaA catalytic triad. Consistent with its higher activity for larger substrates, the Rhodococcus enzyme has a substantially larger active site cavity than DhlA.

An understanding of the catalytic mechanism of dehalogenases has emerged from detailed studies on the haloalkane dehalogenase from X. autotrophicus GJ10. These include determination of X-ray structures of reaction intermediates, site-directed mutagenesis of catalytic residues, and mechanistic and kinetic studies. The enzymecatalysed dehalogenation reaction involves the SN2 displacement of halide by the nucleophilic carboxylate of Asp124 at the active site of the dehalogenase, forming a covalently bound ester intermediate. In the second step the ester intermediate is hydrolyzed by a water molecule that is activated by the diad His289 and Asp260.

During cleavage of the carbon-halogen bond the leaving halide is stabilized by interaction with the NH's of

Trp125 and Trp175. Release of the halide is the last step in the catalytic cycle. Structural analysis of DhlA also suggested ways of engineering enzyme activity. In particular, alteration of the cap domain was found to increase the dehalogenase activity for some substrates. However, attempts to improve the dehalogenase activity for environmentally important compounds such as DCP and TCP have been unsuccessful. Recently, with the development of rapid gene evolution technologies, there is an increasing commercial interest in engineering haloalkane dehalogenases for application in industrial processes, such as waste remediation and chemical synthesis.

Haloalcohol Dehalogenases

The haloalcohol dehalogenase reaction was discovered in 1968 by Castro and Bartnicki in a Flavobacterium isolated on 2,3-dibromo-1-propanol. Since then a number of bacteria have been isolated that produce haloalcohol dehalogenases. These enzymes occur in the degradation pathways of halopropanols and 1,2-dibromoethane. The reaction catalyzed by a haloalcohol dehalogenase involves an intramolecular substitution of a halogen by a hydroxyl group in vic-halohydrins, yielding an epoxide, a proton, and a halide ion.

Haloalcohol dehalogenases have commercial applicability in bioremediation and, more important, in industrial biocatalysis. The stereoselectivity of haloalcohol dehalogenases has allowed the development of microbial processes for the manufacture of optically active compounds, such as epichlorohydrin and glycidol. These compounds find application as chiral intermediates in the production of pharmaceuticals and agrochemicals. Industrial interest in haloalcohol dehalogenases has therefore focused mainly on the group of C3-halohydrins including 1,3-dichloro-2-propanol, 2,3-dichloro-1-propanol, and 3-chloro-1,2-propanediol.

Six haloalcohol dehalogenases have been described to date: the enzymes from Corynebacterium sp. strain N-1074 (HheA and HheB), Arthrobacter sp. strain AD2 (HheBAD2), Agrobacterium sp. (23DCPOHase), Arthrobacter erithii H10a (DehA), and Agrobacterium radiobacter strain AD1 (HheC). Haloalcohol dehalogenases are multimeric proteins composed of 2-4 subunits ranging in size from 28 to 35 kDa. The different dehalogenases are active on a range of C2, C3, and aromatic halohydrins. However, there are some distinct differences in substrate specificity and stereoselectivity between these enzymes. Whereas all haloalcohol dehalogenases described to date are active on 1,3-dichloro-2-propanol, only HheC and 23DCPOHase exhibited considerable activity on 2,3-dichloro-1-propanol.

Insight in the catalytic mechanism of haloalcohol dehalogenases has been hampered by the lack of a three-dimensional structure. Sequence analysis suggests that the overall structure of these enzymes closely resemble that of short-chain dehydrogenases/reductases (SDRs). Using the coordinates of SDR family members as template, a model of a part of the HheC structure was predicted which showed considerable similarity to the X-ray structure of 7a-hydroxysteroid dehydrogenase.

In the model, the position of the essential catalytic residues of 7a-hydroxysteroid dehydrogenase are conserved in HheC. Subsequent sitedirected mutagenesis showed that a triad formed by the amino acid residues Ser132, Tyr145, and Arg149 was essential for catalysis in haloalcohol dehalogenase. The proposed catalytic mechanism for HheC involves proton abstraction from the hydroxyl group by Tyr145, accompanied by a nucleophilic attack of the oxyanion on the adjacent halogen-substituted carbon atom, resulting in ring closure and displacement of halide.

The catalytic serine 132 plays a role in substrate binding by hydrogen bonding to the hydroxyl oxygen, whereas Arg149 may be involved in lowering the pKa of the Tyr145 causing it to be deprotonated. Haloalcohol dehalogenases constitute a unique type of dehalogenating enzymes, since these enzymes do not use a covalent mechanism for catalysis as was found for the well-studied haloalkane dehalogenases.

Monooxygenases

Oxidation of haloalkanes can lead to dehalogenation as a result of the spontaneous elimination of a halide from an unstable geminal halohydrin, which is generated during the oxidation of haloalkanes catalyzed by a monooxygenase. Monooxygenases are multifunctional enzymes that catalyze a range of reactions, and are normally produced to initiate the oxidation of growth-supporting substrates. Organisms producing such enzymes include bacteria that oxidize toluene, phenol, methane, propane, propylene, cumene, isoprene, and ammonia. In addition, it has been shown that these bacteria often can also co-oxidize a variety of halogenated aliphatic compounds. This has led to the study and development of microbiological techniques for bioremediation of halogenated aliphatics that do not support growth of bacteria.

The methane monooxygenase and ammonia monooxygenase produced by methanotrophic and nitrifying bacteria, respectively, have been shown to be active on a number of halopropanes. However the conversion rates are very low and degradation is often incomplete. Methane monooxygenases have been studied in great detail. Dependent on the growth conditions a soluble or a particulate (membrane-bound) methane monooxygenase is produced. Virtually all methanotrophs can express a copper-containing membrane-bound particulate methane monooxygenase (pMMO) with a narrow substrate range. Under conditions of low copper availability some methanotrophs can express a soluble methane monooxygenase (sMMO) that contains two iron atoms in the active site.

The sMMO oxidizes a wider range of halogenated aliphatics than pMMO. The sMMO proteins isolated from Methylococcus capsulatus (Bath) and Methylosinus trichosporium OB3b have been studied extensively (reviewed in and and references therein). The genes encoding the sMMO proteins have been identified and sequenced. sMMO consist of three proteins, a multimeric hydroxylase (251 kDa), a reductase (38 kDa), and a coupling protein (15.5 kDa), all three of which are required for efficient catalysis. The threedimensional structures of the hydroxylase and the coupling protein have been solved by Xray crystallography and NMR-spectroscopy, respectively.

The hydroxylase protein is a $\alpha_2\beta_2\gamma_2$ heterodimer containing a non-heme diiron centre where dioxygen activation and methane hydroxylation occur. The reductase protein provides the electrons for the hydroxylase reaction, where they are used for the activation of O_2. The coupling protein appears to have several regulatory activities. The sMMO active site contains a hydrophobic substrate-binding pocket that can accommodate a wide variety of hydrocarbon substrates besides methane, including saturated and unsaturated, linear, branched, and cyclic hydrocarbons of up to about C8, as well as aromatic, heterocyclic, and chlorinated compounds.

The environmental relevance of sMMO is related to the wide range of reactions that it can catalyze and to its broad substrate range. This makes sMMO a promising enzyme to use for the degradation of recalcitrant environmental pollutants. However, sMMOs are complex enzymes, involving multiple subunits, and requiring a soluble cofactor. The heterologous expression of the hydroxylase protein has also proven to be problematic. Furthermore, conversion of halogenated compounds often leads to the formation of reactive products, such as acylchlorides, that may inactivate the sMMO. Altogether this complicates the development of treatment systems based on the activity of sMMO.

ENGINEERING CATABOLIC PATHWAYS FOR BIOREMEDIATION

The appearance of bacteria that can degrade various man-made chemicals indicated that under selective pressure of environmental pollution, microbial capacity for the degradation of halogenated aliphatics may develop. Genetic analysis of the bacteria involved teaches us a lot about new ways to deal with yet non-degradable recalcitrant pollutants. New catabolic pathways evolve through a natural genetic engineering process. This involves mutation, recombination of sequences, and transfer of DNA segments between different microbial species.

These adaptation mechanisms, which apparently do occur in the environment, can in principle also be exploited under laboratory conditions in order to obtain hybrid organisms, recombined catabolic operons, or modified proteins with desired properties. The ability to mimic and accelerate natural evolution in the laboratory could shave years of the lag period between the introduction of a new synthetic compound into the environment and the development of a microbial capacity that will remove it. A number of approaches can be used to obtain organisms with the desired biodegradative capabilities.

In Vivo Selection

The simplest strategy is to enrich bacteria with new degradative capabilities by sustained selective pressure in chemostats, in batch enrichments, or using plate cultures. This approach does not require detailed information about the degradation mechanism, but it can take considerable time before a stable pure culture is obtained and in many cases no adaptation is found even after years. In vivo selection often depends on the spontaneous transfer of genetic elements encoding catabolic functions, either within a mixed population of bacteria present in the inoculum, or within the pure culture itself.

Natural gene transfer events are the basis of various in vivo design experiments, which are greatly facilitated by the fact that many catabolic pathways for the degradation of halogenated compounds are encoded on mobile DNA elements such as conjugative plasmids or transposons. The continuous culture selection experiments of Kellogg and coworkers is an illustrative example of how the in vivo approach can be used to isolate microorganisms capable of utilizing highly chlorinated and persistent compounds, such as 2,4,5-trichlorophenoxyacetic acid (2,4,5-T), as sole source of carbon and energy.

The synthetic organic compound 2,4,5-T has been released in massive amounts as a herbicide for weed control and was used as a defoliating agent during the Vietnam War (Agent Orange). The natural microflora degrades 2,4,5-T very slowly by co-oxidation, which results in the persistence of this toxic chemical for long periods of time in the environment. Incubation of bacteria harboring a variety of catabolic plasmids in a chemostat yielded after 8 to 10 months of selection on increasing concentrations of 2,4,5-T a pure culture of Burkholderia cepacia that was capable of utilizing 2,4,5-T as a sole carbon and energy source.

The evolution of this catabolic capacity occurred most likely through recruitment of genes present in the chemostat consortium and their integration into the genome. The presence of insertion sequences adjacent to the gene clusters involved in 2,4,5-T metabolism supports this idea.

In vivo selection has also proven to be useful for the isolation of mutants of key enzymes that constitute barriers for the metabolism of certain xenobiotics. The pioneering studies of Clarke and coworkers on the Pseudomonas aeruginosa amidase describe one of the classical experiments in laboratory evolution of enzymes with novel activities. P. aeruginosa can use acetamide as a growth substrate, since the amidase can hydrolyze this compound to ammonia and acetate, which can be used as a source of nitrogen and carbon, respectively.

Using a variety of acetamide analogues it was possible to select many mutant P. aeruginosa strains that produce amidases with altered substrate specificities, which enabled the mutants to grow on aliphatic amides that cannot be utilized by the wild-type strain. In vivo selection was also successfully applied to obtain mutants of haloalkane dehalogenase DhlA with different substrate specificity. Wild-type DhlA slowly dehalogenated 1-chlorohexane, which hampered the mineralization of this compound by the recombinant strain Pseudomonas putida GJ31(pPJ20) expressing the dhlA gene. A few months of selection in batch culture yielded a number of mutant strains that could grow on 1-chlorohexane. Six different variants of DhlA were obtained with improved activity on 1-chlorohexane. All mutations were located in the cap domain, indicating an important role of this domain in determining substrate specificity of haloalkane dehalogenases.

In Vitro Construction of Catabolic Pathways

The possibility of using gene technology allows the rational combination of useful catabolic genes to create hybrid metabolic pathways for improving biodegradation of recalcitrant pollutants. This requires detailed genetic and biochemical information on the degradation pathways of xenobiotics. In the past decades considerable progress has been made in exploring catabolic genes, especially for halogenated aromatic compounds. Lehrbach et al. were the first who reported of an in vitro constructed catabolic pathway.

Introduction of three catabolic genes cloned on a broad host range plasmid expanded the catabolic range of Pseudomonas sp. B13 to include 3-chlorobenzoate, 4-chlorobenzoate, and 3,5-dichlorobenzoate. By combining genes from different origins they created hybrid pathways that are able to eliminate combinations of synthetic compounds, such as mixtures of chloro-benzoates and alkyl-benzoates. Whereas the separate compounds are easy biodegradable, these compounds persist when they are present as mixtures due to the induction of incompatible pathways. The engineering of bacteria for the degradation of polyhalogenated compounds is another example that demonstrates the value of recombinant DNA technology for constructing bacteria with broad metabolic capacities.

These compounds are generally very resistant to biodegradation and degradation typically requires multiple mechanisms of dehalogenation. Polyhalogenated compounds are most often partly metabolized by anaerobic bacteria via reductive dehalogenation reactions. The products of these reductive dehalogenations are typically substrates for bacterial oxygenases. This necessitates the sequential use of anaerobic and aerobic bacteria for the degradation of polyhalogenated compounds, which appears to be cumbersome from an engineering point of view. To circumvent this, both activities have been combined into a single organism.

PROTEIN DESIGN METHODS

Protein design methods can be used to modify the selectivity and activity of catabolic enzymes, which offers opportunities for enhancing the degradation of recalcitrant compounds. Although there are many ways to evolve proteins, the strategy of most methods is to create variants and select those that perform best. In general two protein engineering strategies can be distinguished: existing enzymes can either be tailored by rational (structurebased) redesign, or by applying directed evolution methods. The latter method involves the construction of large libraries of variants using random mutagenesis or gene recombination, followed by selection or screening to identify the mutants with desired traits.

Rational Design

Rational design of proteins relies on detailed knowledge of protein structure, function and mechanism, and on molecular modeling to predict favorable amino acid changes. A small number of promising modifications are introduced into the gene using site-directed mutagenesis, and the effects are tested. Rational approaches have been used recently to engineer properties of enzymes involved in the degradation of xenobiotics. For example, guided by structural data of reaction intermediates of haloalkane dehalogenase from X. autotrophicus GJ10, bulky amino acids lining the active site cavity were replaced by smaller residues increasing the volume of the active site. The resulting variants showed improved activity on 1,6-dichlorohexane compared to the wild type.

In the absence of structural information, sequence comparison of related enzymes could also facilitate the choice for a particular mutational site. This was, for instance, applied to influence the substrate specificity of biphenyl dioxygenases. Based on the modular nature of many catabolic enzymes more drastic changes in protein function can be obtained by the construction of hybrid enzymes. Enhanced conversion rates on trichloroethylene, for instance, were obtained with hybrid enzymes created from a toluene and biphenyl dioxygenase.

Recently, a combinatorial approach was developed for the creation of hybrid enzymes. This technique, designated ITCHY (Incremental Truncation to Create Hybrid enzymes), allows the combination of protein domains with low or no appreciable degree of homology. Although rational design has proven fruitful, our insufficient understanding of enzyme-substrate interactions that contribute to catalysis impedes this approach in many cases, so that variants frequently fail to show the desired improvements. Currently, evolutionary protein design methods are the most effective approaches for improving enzyme function. These methods have been developed to such an extent that improved enzymes can be obtained on a reasonable timescale, and require little or no information regarding protein structure and function, as will be discussed below.

Random Point Mutagenesis

Random point mutagenesis of the gene of interest is typically performed by errorprone PCR, although mutator strains and exposure to chemical mutagens have also been used. A library of mutant enzymes, containing one or two amino acid substitutions, is created and searched with an appropriate selection or screen to identify functionally improved variants. The winner of the first round is then subjected to additional iterative rounds of random mutagenesis and selection. This process results in the accumulation of single mutations over many generations. Sequential rounds of random mutagenesis is a simple and effective strategy that has been successively used to evolve proteins with enhanced activity, altered substrate specificity, thermostability, solvent tolerance, and improved enantioselectivity.

A large family of enzymes that might find applications in areas ranging from chemical synthesis to environmental remediation are cytochrome P450 monooxygenases.

Interest in these enzymes is based on the ability to catalyze the insertion of oxygen into a wide variety of compounds, including many xenobiotics. Using random mutagenesis and a fluorescent whole-cell screen, Arnold and coworkers identified several variants of cytochrome P450CAM monooxygenase from P. putida with high activity against naphthalene compared to the native enzyme.

Iterated random point mutagenesis proved to be very successful in cases where most likely several sequence solutions exist and the obtained mutations are additive, for example when improving the stability of an enzyme. To evolve complex systems such as an operon, or in case that a new protein function requires the simultaneous introduction of multiple amino acid substitutions, random mutagenesis will fail. For example, despite extensive efforts no improved variants of the dibenzothiophene monooxygenase (DBT-MO) were obtained after random point mutagenesis of dszC genes. However, in vitro recombination of two homologous dszC genes yielded variants with higher activity and broader substrate range, most likely because recombination allows more diverse permutations of multiple beneficial mutations than random mutagenesis alone.

In Vitro Recombination

An important aspect of natural evolution that is lacking in the mutagenesis methods described above is the ability to recombine mutations from individual genes, akin to natural sexual recombination. Recombination followed by selection combines useful mutations and simultaneously flushes out harmful mutations. The view of recombination as the key accelerator of evolution has inspired the invention of in vitro recombination methods. The most widely used method for in vitro recombination, pioneered by Stemmer, is based on random fragmentation of a pool of related DNA sequences and reassembly of the fragments into full-length genes in a primerless DNA amplification reaction (gene shuffling).

During this process random point mutations are introduced at a controllable rate. Multiple cycles of fragment reassembly results in crossovers in regions with sequence homology due to template switching creating novel combinations of mutations. Subsequent expression of the chimeric sequences followed by an effective screen or selection identifies improved variants. A pool of the best clones can then be subjected to additional rounds of DNA shuffling and selection or screening, until the evolved sequence encodes the desired function.

Originally, DNA shuffling has been used for the recombination of mutations present in single gene sequences. However, the most powerful advantage of DNA shuffling is that it allows recombination of multiple parental gene sequences, which dramatically accelerates evolution of protein function. This has been nicely demonstrated for subtilisin. Due to their importance in the detergent industry, subtilisins have frequently been targeted in the past 30 years for improvements using rational design and random mutagenesis. Although the improvements have been notable, they are often at the expense of other properties that are critically required for industrial implementation such as thermostability, solvent stability and pH dependence.

DNA shuffling of 26 protease genes obtained from Bacillus yielded several chimeric proteases with multiple improved properties, and new combinations of properties not exhibited by any of the parent enzymes. DNA shuffling has been applied to a variety of enzymes to improve protein folding, substrate specificity and enzyme activity under diverse physical and chemical conditions. In addition to improving single gene sequences, DNA shuffling can also address more complex problems, such as optimizing an entire biodegradative pathway. The arsenate operon of

Staphylococcus aureus, for example, was evolved by three rounds of DNA shuffling. Cells expressing the optimized operon were able to grow on 0.5 M arsenate, corresponding to a 40-fold increase in resistance. Furthermore, whole cells can be improved by recombining entire genomes.

Although application of evolutionary protein design methods has been mainly focused on improving enzymes for industrial biocatalysis, there is an increasing interest in evolving enzymes for bioremediation.

BIOLOGICAL PARTICLES IN ATMOSPHERIC CLOUDS

Biological particles may, like aerosol particles of other origin, influence cloud formation and precipitation processes via the following paths: (i) the phase change from vapour to liquid, (ii) the acceleration of coalescence by large particles; and (iii) the phase change from vapour or liquid to ice. The other two processes depend on more specific properties, as will be discussed. It may be pointed out that from what is known at this time, ice nucleation by bio-aerosols is expected to have the greatest potential for influencing cloud evolution.

ICE FORMATION IN CLOUDS

It is a well-established fact in meteorology that the presence of ice particles in clouds is a major factor in the formation of precipitation, in the radiative properties of clouds and in chemical interactions within clouds. It is also clearly established that a substantial fraction of clouds in the troposphere is composed totally or partially of ice particles. A major part of global precipitation, whether it arrives at the Earth's surface as rain or as some form of ice, has gone through some ice process. The principal reasons for this are that clouds easily extend to the colder temperatures of the atmosphere and in a situation where both liquid and solid cloud particles exist, the latter are favoured to grow to precipitation sizes.

Yet, while the importance of the ice phase in clouds is clear, major uncertainties still exist about the factors controlling the origins and abundance of ice particles in clouds. Observations show that ice concentrations are much lower and vary over a much larger range than cloud droplets. From the lowest measurement limit of perhaps $1 m^{-3}$, ice concentrations can reach up to $10^6 m^{-3}$ or more. In comparison, liquid droplet concentrations range from about 10^7 to $10^{11} m^{-3}$. The question of ice nucleation in clouds is further complicated by the fact that there are no general theories - in a comparable way to the roles of solubility and size for CCN activity - to indicate what characteristics determine ice nucleating activity. Some factors are known to be relevant and will be mentioned later.

A rather significant consequence of the relatively low number concentration of ice nuclei in the atmosphere is that ice does not appear in clouds with the same sharp threshold as is observed for condensation. Essentially all cloud droplets form within a few tens of meters at the base of a cumulus cloud. In contrast, ice particles originate in a more distributed fashion and most clouds are supercooled for some portion of their evolution, leading to 'mixed-phase clouds' in which supercooled droplets are intermingled with some ice particles.

Under such conditions, the lower vapour pressure of ice favours a transfer of water mass from the droplets to the ice particles, with the consequent growth of ice, and diminution of cloud water mass. This is known as the Wegener-Bergeron-Findeisen process. Once the ice particles acquire enough mass to fall at appreciable velocities, they also collect droplets by collisions, a process called riming. These are the basic elements of precipitation formation in the vast majority of clouds, albeit they are not the whole story; the growth of various ice crystal shapes, the intricacies of the riming process leading to graupel and possibly to hailstones, the aggregation of crystals into snowflakes, and so on.

This is also a general feature of theoretical predictions. The manifestation of this trend in clouds is that often the concentration (number per volume of air) of ice particles increases with decreasing temperatures. In such cases, the ice particles that start at the highest temperatures below 0°C have the longest possible growth times and the greatest chance to develop to precipitation sizes. This is one of the reasons for the special interest in biogenic ice nuclei, since, as will be shown, they are known to be capable of nucleating ice at those high temperatures. However, the situation is more complex. It has not been possible up to this time to demonstrate a clear agreement between ice nucleus measurements and ice concentrations in clouds. Reasons for this are many.

Most importantly, both types of measurements are exceedingly difficult to make. Secondly, it is known that ice can originate from secondary processes involving interactions among ice particles or with water droplets. By contrast, ice nucleation is termed the primary process of ice generation. As a result, the expectation of monotonic increase in ice concentrations with decreasing temperatures that would result from ice nucleation is of limited validity, and the limits are not well known. The overall result is that the prediction of ice concentrations in clouds, from either direct ice nucleus measurements or from other characteristics of the aerosol, is currently possible to rather limited extents. This situation is further clarified in the following.

Ensemble analyses of ice concentrations in stratiform and frontal clouds showed no relation between temperature and the concentrations of ice crystals down to sizes of 125 μm. Many of these observations cannot be explained by known secondary ice formation processes. The discrepancy between these observations and the expectations based on ice formation by ice nuclei may be partly due to ice crystal redistribution in the clouds, but may also indicate the possibility that primary and/or secondary ice formation mechanisms not yet identified are responsible.

Some investigators suggest that evaporation of cloud droplets is associated with the formation of ice. No mechanism for this has been identified so far, and present techniques of measuring ice nucleation by aerosols do not detect this process.

ICE NUCLEATION

Ice can form in the atmosphere either from the vapour or from the liquid phases of water - these are distinguished as "deposition" and "freezing". The ice nucleation pathway of interest here is catalyzed by some particle - heterogeneous nucleation. At temperatures approaching -40°C, supercooled liquid will freeze spontaneously, by homogeneous freezing nucleation. This latter process is well studied. It is time-dependent, with the rate rising extremely rapidly near -35°C, is slightly modified by dissolved substances in the water, and for a population of droplets means that freezing is most likely in the larger droplets. The latter arises from the fact that the probability of homogeneous freezing per unit volume and per unit time is the fundamental quantity governing the process.

This was demonstrated in recent work for water droplets larger than a few microns in diameter suggested that homogeneous freezing may occur preferentially close to the surface-air interface of drops. At least for droplets smaller than a few micron in diameter existing data are not of sufficient quantitative accuracy to differentiate whether surface or volumetric nucleation rates dominate. Heterogeneous ice nucleation can take on further variations in clouds, beyond the basic ones of deposition and freezing. Laboratory results showed that the prior history of how a particle gets into the liquid makes a difference; this led to identifying three modes of freezing nucleation: immersion-, condensation- and contact-freezing.

Furthermore, nucleation activity has been found to be influenced in some cases by pre-activation or prior exposure to low temperatures and/or humidity conditions, by memory (hysteresis) effects, UV radiation, and electric fields among other things. On the other hand, repeated cycles of freezing and melting in the laboratory lead to the conclusion that most ice nucleation is due to robust features that cause freezing at close to the same temperature over and over.

Theoretical treatments lead to estimates of the sizes of the surface sites responsible for nucleation. That size decreases with temperature, from the order of 0.01 μm for freezing a few degrees below 0°C to a tenth of that for temperatures near -20°C. However, there is no empirical confirmation for these figures. Beyond size, both theory and ever, observations indicate that the nucleating sites are likely to possess crystallographic arrangements of molecules similar to that of water. Silver iodide is one of the sub- substances that bear out this fact. It is also believed that steps, cavities and other surface stances features of crystalline materials might enhance their potential for ice nucleation. Hydrogen bonding to the ice lattice is likely to be a factor for organic substances.

During the past decades, the knowledge about the nature and number concentration of atmospheric ice nuclei has significantly been improved by the application of new instruments and experimental methods in field measurements and laboratory studies. There are two different ways to study the nature of ice nuclei in the atmosphere. The first is to look at particles at the centres of snow crystals or in their residues.

A variation of this is to collect small ice crystals following nucleation in a laboratory cloud chamber or field instrument and then analysing the residual particles by electron microscopy particle mass spectrometry. Second, relationships can be sought between aerosol composition, as determined by any of the variety of techniques available, and the abundance of ice in clouds that form in the sampled air mass.

For the first approach of direct analysis of residual particles from evaporating ice crystals care has to be taken that after the nucleation event the sampled ice crystals grew mainly by vapour diffusion and did not collect further aerosol particles e.g. by scavenging or riming processes. It should be noted that even if pristine, freshly nucleated ice crystals are sampled to avoid the effect of secondary particle collection, there is still the chance that non-detectable trace mixtures of other materials on a larger carrier particle may be responsible for ice nucleation.

The correlation approach is affected by the fact that ice nuclei constitute a very small fraction of the total aerosol population in the atmosphere, so that a significant parallel between the aerosol composition and ice nucleation can be established only if there is a dominant and well identifiable source of aerosol that contributes to the observed ice formation.

Field measurements of atmospheric ice nuclei have been made with a variety of different instruments. These measurements yield ice nuclei spectra: concentration as a function of temperature and/or some measure of supersaturation (with respect to water or ice). Fundamentally, all of these instruments attempt to mimic cloud processes by creating equivalent conditions within a chamber filled with outside air.

The degree of realism of these simulations varies greatly and may be limited to one, or a few of the possible modes of ice nucleation activity. At temperatures above -10°C, the low concentration of atmospheric ice nuclei makes it practically impossible to produce meaningful measurements in instruments of sizes that can be deployed in the field.

Both inorganic and organic particles have been identified as atmospherically relevant ice nuclei. Mineral dusts are a general category of inorganic particles that are a reported and sometimes strong source of ice nuclei to the atmosphere. Biological particles, including certain bacteria, pollen, and decayed organic material have also been identified as ice nuclei. At present, there exists only some indirect and modest direct evidence that ice nuclei predict ice formation in clouds. The indirect evidence comes from observing the onset of ice formation in clouds or cloud regions where the action of primary ice nucleation formation processes is expected to be isolated.

CLOUD MODELLING

In most cloud models the number of ice nuclei which are available to initiate the formation of primary ice particles in supercooled liquid clouds is calculated with simple equations mainly as function of the temperature and/or the ice supersaturation that serve as surrogates for explicitly resolving deposition and freezing mechanisms. These are sometimes supplemented with a formulation for contact-freezing nucleation. Such equations were, for example, deduced by Cotton et al. as a combination of earlier work partially based on measurements of atmospheric ice nuclei using a static filter approach by Meyers et al. using measurements of ice nuclei concentrations with novel continuous flow diffusion chambers or by Cooper using measurements of ice particle number concentrations in the top layers of clouds where secondary ice nucleation processes had only a minor or negligible contribution to actual ice particle concentration. It is one of the major challenges of cloud physics to improve these equations and to formulate the formation of ice in cloud models also as a function of aerosol particles.

Although biological particles were found on mineral dust particles in the atmosphere and in cloud water samples, there is no clear evidence that they play a role in the formation of ice in clouds. Part of the problem may be related to the inability of the present IN measurement methods to identify biological particles as ice nuclei in cloud ice crystals and separating them from the inorganic substances. Furthermore, even if found inside ice crystals it is not yet clear what their importance is in the development of the clouds. This is partly because their concentrations may be too low, or that their effectiveness is expressed at temperatures not too relevant for the subsequent growth of the clouds.

The impact of biological particles on cloud development and initiation of precipitation could be very significant if they act as large or giant CCN and also as IN. The presence of such particles in an ascending air parcel of a convective cloud could form a few large droplets that would grow rapidly to sizes larger than 25 μm, before reaching the -4 to -8°C temperature levels. At these levels the large droplets could play a role in the Hallett-Mossop mechanism by riming with existing ice crystals. Furthermore, being effective immersion IN, these bio-aerosols can freeze the larger droplets.

REFERENCES

Gessner, M.O. and Newell, S.Y., "Biomass, growth rate, and production of filamentous fungi in plant litter", *Pages 390-408 in: Manual of Environmental Microbiology*, Second Edition, ASM Press, Washington, D.C., 2002.

Hanson, J.R., J. L. Macalady, D. Harris and K.M. Scow, "Linking toluene degradation with specific microbial populations in soil", *Applied and Environmental Microbiology*, 1999.

Jetten et al, "Microbiology and application of the anaerobic ammonium oxidation (anammox) process", *Environmental Biotechnology*, 2001.

10

Industrial Microbiology

Industrial microbiology encompasses activities like production of biocontrol agents, inoculants used as biofertilizers, etc. The activities in industrial microbiology begin with the isolation of microorganisms from nature, their screening for product formation, improvement of product yields, maintenance of cultures, mass culture using bioreactors, and usually end with the recovery of products and their purification.

PRODUCTS AND PROCESSES OF INDUSTRIAL MICROBIOLOGY

Industrial microbiology also deals with products associated with the food and dairy industries, with the prevention of deterioration of processed or manufactured goods and with waste disposal systems. Many industrial microbiologists/ biotechnologists are responsible for the discovery, development or implementation of certain processes and the quality of resultant products. Such products and processes include:

— *Antibiotics/Antimicrobials*. Both natural and chemically enhanced microbial products can be used to control human, animal and plant diseases. Using traditional genetics or recombinant DNA techniques, the microorganism can be modified to improve the yield or action of antibiotics and other antimicrobial agents. New research directions are aimed at discovering microbial metabolites with pharmacological activities useful in the treatment of hypertension, obesity, coronary heart disease, cancer and inflammation.

— *Vaccines.* Vaccines are essential to protect humans and animals from microbial diseases. Recombinant DNA technology has allowed the production of novel vaccines that offer protection without the risk of infection (e.g. hepatitis B vaccine).

— *Health-Care Products.* The development and production of diagnostic assays that utilise monoclonal antibody or DNA probe technology are essential in the manufacture of health-care products such as rapid tests for strep throat, pregnancy or AIDS. Other examples include the use of microbial cells to produce human or animal biologicals such as insulin, growth hormone and antibodies. The industrial microbiologist/ biotechnologist may screen new microbial sources (e.g., marine microorganisms) for their ability to produce new pharmaceuticals.

— *Foods/Beverages produced by microbial activity.* Yogurt, cheese, chocolate, butter, pickles, sauerkraut, soy sauce, food supplements (such as vitamins and amino acids), food thickeners (produced from microbial polysaccharides), alcohol (beer, whiskeys, wines) and silage for animals are all products of microbial activity. The industrial microbiologist/biotechnologist may be involved in producing concentrated microbial inocula for fermentations or the maintenance of fermentation systems utilised in production facilities.

— *Foods/Beverages cured or improved by microbial activity.* Production of coffee, tea, cocoa, summer sausage, vanilla, cheese, olives and tobacco all require microbial activity.

— *Food flavoring agents and preservatives.* Organic acids such as citric, malic and ascorbic acids and monosodium glutamate are microbial products commonly used in foods.

— *Foods*. Mushrooms, truffles and some red and green algae are consumed directly. Yeasts are used as food supplements for humans and animals.

— *Agriculture*. Conventional, recombinant DNA and monoclonal antibody techniques are used to improve microbial inoculants which serve as fertilizer supplements by fixing atmospheric nitrogen and as plant pest controls.

— *Enzymes*. Industrial applications of enzymes include the production of cheese, the clarification of apple juice, the development of more efficient laundry detergents, pulp and paper production and the treatment of sewage. These processes have been dramatically enhanced by the use of recombinant DNA techniques to design enzymes of increased activity, stability and specificity. Carbohydrates. Molecular sieves (e.g., dextran) for purification/separation processes; and thickening agents (e.g., xanthan) that are stable at high temperature are examples of microbial carbohydrates. The latter are used for secondary oil recovery in oil fields, lubricants in drilling oil wells, gelling agents in foods and thickeners in both paints and foods.

— *Carbohydrates*. Molecular sieves (e.g., dextran) for purification/separation processes; and thickening agents (e.g., xanthan) that are stable at high temperature are examples of microbial carbohydrates. The latter are used for secondary oil recovery in oil fields, lubricants in drilling oil wells, gelling agents in foods and thickeners in both paints and foods.

— *Organic chemicals*. Compounds such as acetone, methanol, butanol and ethanol all have multiple applications in industrial settings. Microbes will increasingly be used to supplant or replace those processes which rely on petroleum/natural gas for the production of these compounds.

— *Oil recovery/mining*. Oil recovery may be facilitated by the development of unique bacteria which produce a surfactant that forces trapped oil out of rocks. Extraction of minerals from low-grade ores is enhanced by some bacteria (microbial leaching). In addition, selective binding of metals by biohydrometallurgical processes is important in recycling of metals such as silver and uranium.

— *Contamination control*. Development of assays to detect microbial contaminants in food; evaluation of natural or synthetic agents for the prevention of disease, deterioration or spoilage; determination of minute quantities of vitamins or amino acids in food samples; development of preservatives for control of food spoilage; and development of procedures for control of deterioration in cosmetics, steel, rubber, textiles, paint and petroleum products rely on extensive knowledge of industrial microbiology and biotechnology.

— *Waste and wastewater management*. Isolating or developing microbial strains capable of degrading and detoxifying hydrocarbon and halogenated hydrocarbon waste (for example organophosphates, acetylcholinesterase inhibitors) of industrial, agricultural or military origin is essential in waste management.

— *Environmental science*. Understanding the role of microorganisms in the degradation and transformation of chemicals in the environment is a key factor in assessing the environmental safety of new and existing chemicals. Engineering microorganisms may solve contamination and recycling process problems.

Nanobiotechnology

Small size, unique physical structures, and electronic

properties have sparked an explosion of research in nanobiotechnology. "Carbon buckyballs and nanotubes are small enough to move inside the body serving as effective carriers of active ingredients," says the analyst. "For instance ABI 007 is a successfully tested nanocarrier that transports established anti-cancer drugs. Minute nanoparticles such as quantum dots may serve as imaging agents, revolutionising conventional diagnostic imaging."

The application of extremophiles in industrial processes is set to usher in a new era in biotechnology. Ground breaking research into their unique ability to survive under harsh physical conditions creates numerous opportunities in bioprocessing and environmental biotechnology. Also, the potential of secondary metabolites derived from extremophiles in replacing certain industrial bioproducts is stimulating research in this area.

Chemical Industry

Chemical industry includes those industries involved in the production of petrochemicals, agrochemicals, pharmaceuticals, polymers, paints, oleochemicals etc. Chemical processes are used, including chemical reactions to form new substances, separations based on properties such as solubility or ionic charge, and distillations, in addition to transformations by heating and other methods.

Chemical industries involve the processing of, or change in, raw materials obtained by mining, and agriculture among other supply sources, into materials and substances that are useful on their own, or in other industries. The food-processing industries are generally not included in the term "chemical industry".

Fossil Fuels

Fossil fuels, also known as mineral fuels, are hydrocarbon-containing natural resources such as coal, petroleum and natural gas. The utilisation of fossil fuels has fueled industrial development and largely supplanted water driven mills, as well as the burning of wood or peat for heat. When generating electricity, energy from the combustion of fossil fuels is often used to power a turbine. Older generators often used steam generated by the burning of the fuel to turn the turbine, but in newer power plants the gases produced by burning of the fuel turn a gas turbine directly.

The burning of fossil fuels by humans is their major source of emissions of carbon dioxide which is one of the greenhouse gases that is believed to contribute to global warming. A small amount of hydrocarbon-based fuels are biofuels which are derived from atmospheric carbon dioxide and thus do not increase the carbon dioxide in the atmosphere.

Industrial Dynamics

Biotechnology holds clear promises of becoming the next general purpose technology. Biotechnology can be considered a platform technology comprising a large array of sub-technologies based on a series of key scientific advances in the second half of the 20th century that established new insights into the biological basis of life. Biotechnology develops and builds on knowledge and capabilities in a diverse set of disciplines, especially molecular biology, but also chemistry, biochemical engineering, microbiology, computing and programming.

Biotechnology is "technology", not just "science", because it represents huge commercial opportunities for innovation. But, contrary to other technologies, the activities underlying the development of biotechnology strongly overlap with those underlying fundamental science. Biotechnology thus both reflects a new scientific and technological paradigm; a context which until now especially has had substantial impacts on the organisation and dynamics of the pharmaceutical industry.

But biotechnology can also be conceived as an "industry", a collection of firms that primarily specialise in research and development of biotechnology. Especially in the USA, a virtual bubble of entrepreneurial energy fed into establishing dedicated biotechnology firms (DBFs) during the 1980s and 1990s, founded by university scientists and based on university science. More than elsewhere in the world institutional changes were implemented in the USA both in regard to intellectual property rights (IPR) in life sciences and financing of purely science-based firms, mobilising high-powered incentives among university scientists to "go commercial".

Incumbents in pharmaceuticals have had to acquire and assimilate biotechnology capabilities and to engage in cooperative relations with DBFs, universities and other research institutions in order to survive, and the same seems slowly to become the case, even if at a much smaller scale, in these and other industries. This development is likely to come to resemble the diffusion of information and communication technology (ICT) during the last 20 years, where other industries than the core ICT industries have come to account for an increasing share of overall ICT innovations.

From its embryonic beginnings the biotech industry is now about 30 years old. However, as Maryann Feldman points out in this issue, "[b]iotech is still at an early stage of development and there are many competing hypotheses about its future development". This issue presents five papers representing cutting-edge research on the industrial dynamics of biotechnology which in numerous ways challenge conventional wisdom.

By the year 2025, it's estimated in a recent study that the market for biotechnology-related products in the European Union will be worth ECU250 billion (US$278 billion) – equivalent to around three million jobs. 70% of these jobs will be in the agricultural and food sectors. Already in the

Netherlands, 20% of consumer spending and 18% of the workforce are in the agro-food industries.

A growing area of industrial biotechnology is the development of products and techniques for cleaning up pollution caused by agriculture, industry or urbanisation – this is known as environmental biotechnology or 'bioremediation'.

Industrial Fermentations

Many industrial biotechnology products, such as beer, antibiotics or enzymes, are produced by growing microorganisms, under clean, highly controlled conditions in containers called fermenters. Most fermenters are stainless steel or glass tanks containing anywhere from 10 to 10000 litres of liquid. The liquid contains all of the nutrients that living cells need to grow – carbon and nitrogen sources, vitamins, minerals, phosphorous and oxygen. The concentration of all of the nutrients and oxygen and the temperature are all controlled by computer.

Cells from other organisms – plants or animals – can also be grown in this way, and used to produce large amounts of valuable products. A valuable purple dye called shikonin is produced in this way in Japan from a plant that was almost extinct: the dye was originally extracted from the roots of 7-year-old plants, but can now be harvested from cells in culture after just 20 days.

Substrates Used in Industrial Fermentation.

Carbon Sources

— Carbohydrates are traditional energy sources in the fermentation industry.

— Pure glucose or sucrose can seldom be used as the sole carbon source, except in processes, which demand exact fermentation control as they are very expensive media.

— Molasses, a byproduct of sugar production, is one of the cheapest sources of carbohydrate. Besides a large amount of sugar, molasses contains nitrogenous substances, vitamins, and trace elements. However, the composition of molasses varies depending on the raw material used for sugar production.

— Malt extract, an aqueous extract of malted barley, is an excellent substrate for many fungi, yeasts, and actinomycetes. Dry malt extract consists of about 90-92% carbohydrates, and is composed of hexoses (glucose, fructose), disaccrides (maltose, sucrose), trisaccharides (maltotriose), and dextrins. Nitrogenous substances present in malt extract include proteins, peptides, amino acid, purines, pyrimidines, and vitamins. The amino acids composition of different malt extracts varies according to the grain used, but proline always makes up about 50% of the total amino acids present. Culture media containing malt extract must be carefully sterilised. When overheating occurs, the Maillard reaction results, due to the low pH value and the high proportion of reducing sugar. In this conversion, the amino groups of amines, amino acid (especially lysine), or proteins react with the carbonyl groups of reducing sugars, aldehydes or ketones, which results in the formation of brown condensation products. These reaction products are not suitable substrates for microorganisms. The Maillard reaction is one of the main causes of damage to culture media during heat sterilisation, resulting in considerably reduced yields. pH (10% solution) = 5.5 ·

— Starch and dextrins can be directly metabolised as carbon sources by amylase-producing organisms. In addition to glucose syrup, which is frequently used as a fermentation substrate, starch has become more important as a substrate for ethanol fermentation.

— Sulphite waste liquors, sugar-containing waste products of the paper industry, which have a dry weight of 9-13%, are primarily used in the cultivation of yeasts. Sulphite liquors from coniferous trees have a total sugar content of 2-3%, and 80% of the sugars are hexoses (glucosc, man-nosc galactose), the others being pentoses (xylose arabinose). Sulphite liquors from deciduous contain mainly pentose sugars.

— Because of its wide availability and low cost, cellulose is being extensively studied as a substrate for conversion to sugar or alcohol. It is usually not possible to use cellulose directly as a carbon source, so it must first be hydrolysed chemically or enzymatically. The sugar syrup formed from cellulose hydrolysis has been used for ethanol fermentation, and the fermentative production of butanol, acetone, and isopropanol is also being considered. Work is in progress to develop one-step processes for direct conversion of cellulose to ethanol, using fermentative organisms which produce cellulases.

— Whey, a byproduct of the dairy industry, is produced annually on a world-wide basis to the amount of 74 million tons (containing 1.2 million tons of lactose and 0.2 million tons of milk protein). Only about 56% of this product is used for human or animal feed. The lactose is used primarily for the production of ethanol or singlecell protein, but also in the production of xanthan gum, vitamin B_{12}, 2,3-butandiol, lactic acid, and gibberellic acid. Because of storage and transportation costs, whey is often not economical as a substrate.

— Animal fats such as lard and animal and plant oils are readily utilised by some microorganisms, but are generally added as supplemental substrates rather than as the sole fermentable carbon source. For instance, in certain antibiotic fermentations, soy, palm, and olive oils are used.

— With respect to its carbon content, methanol is the cheapest fermentation substrate, but it can be metabolised by only a few bacteria and yeasts. Methanol has commonly been used as a substrate for single cell protein production. Research has been carried out on processes for the production of glutamic acid, serine, and vitamin B_{12} using methanol as the sole carbon source or as a cosubstrate.

— Ethanol is available in ample supply from the fermentation of either saccharified starch or cellulose, and can be metabolised by many microorganisms as the sole carbon source or as a co-substrate. Acetic acid, for instance, is presently made by the oxidation of ethanol. At the present time, the cost of ethanol is too high to make it utilisable as a general industrial carbon source.

— Alkanes with a chain length of C_{12} to C_{18} are readily metabolised by many microorganisms. The use of alkanes as an alternative to carbohydrates depends on the price of petroleum.

Nitrogen Sources

Many large-scale processes utilise ammonium salts, urea, or gaseous ammonia as nitrogen source. A nitrogen source, which is efficiently metabolised, is corn steep liquor, which is formed during starch production from corn. The concentrated extract (about 4% nitrogen) contains numerous amino acids, such as alanine, arginine, glutamic acid, isoleucine, threonine, valine, phenylalanine, methionine, and cystine. The sugar present in corn steep liquor becomes largely converted to lactic acid (9-20%) by lactic acid bacteria.

— Yeast extracts are excellent substrates for many microorganisms. They are produced from baker's yeast through autolysis at 50-55%C through plasmolysis in the presence of high concentrations of NaCl. Yeast extract contains amino acids and peptides, water-soluble vitamin, and carbohydrates. The glycogen and trehalose of yeast cells are hydrolysed to glucose during yeast extract production. The composition of yeast extract varies partly because the substrates used for yeast cultivation affect the quality of the yeast extract.

— Peptones (protein hydrolysates) can be utilised by many microorganisms but they are relatively expensive for industrial application.

Sources of peptones include meat, casein, gelatin, keratin, peanut seeds, soy meal, cotton seeds, and sunflower seeds. Peptone composition varies depending upon its origin. For instance, peptone from gelatin is rich in proline and hydroxyproline, but has almost no sulphur-containing amino acids. On the other hand, peptone from keratin has a large proportion of proline and cystine, but lacks lysine. Peptones of plant origin (soy peptone, cottonseed peptone) have large proportions of carbohydrates. The end product is also influenced by the type of hydrolysis, whether acid or enzymatic, especially in regard to its tryptophan content.

— Soy meal, the residue from soybeans after the extraction of soybean oil, is a complex substrate. Analysis shows a protein content of 50%, a carbohydrate content of 30% (sucrose, stachyose-, raffinose, arabinoglucan, arabinan, and acidic polysaccharides), 1% residual fat, and 1.8% lecithin. Soy meal is frequently used in, antibiotic fermentations; catabolite regulation does not occur because of the slow catabolism of this complex mixture.

Enzymes

Many of the life processes that biotechnology can make use of are chemical reactions that are controlled by enzymes, which are proteins that act as catalysts to speed up and direct the reactions. Some biotechnological processes use whole organisms (e.g. yeasts to make wine), while others use enzymes that have been purified from the organism that makes them (e.g. enzymes in detergents to dissolve grease and protein).

Enzymes are already widely used by industry – the current global market is estimated at around US$1.5 billion, and set to grow by 5–10% per year. The detergent, food and starch processing industries still account for 75% of 'bulk' enzyme use (mostly hydrolases such as proteases, amylases, lipases and cellulases). 'Specialty' enzymes now account for around 10% of the market and are finding increasing uses in the development of new drugs and antibiotics, as well as medical diagnostics and analytical uses (e.g., for monitoring food safety).

About 60% of enzymes are the product of modern biotechnology, and new uses are being developed for environmental, food, agricultural and textile industry uses. A new class of sugars called isomalto-oligosaccharides is being produced by enzymes called glucosyl transferases. Isomalto-oligosaccharides have potential commercial applications in the food and health industries as non-digestible carbohydrate (bulking agent), an agent that prevents tooth decay, and to prevent baked goods going stale.

Cellulases are complexes of enzymes which, together, break down the cellulose polymers of plant tissue. Cellulases produced by a fungus called Trichoderma are used for 'stonewashing' jeans – but their activity has to be carefully controlled to get the desired effect without destroying the cotton material. Changing the relative proportions of the enzymes in the cellulase complex creates different effects: true 'designer jeans'.

Enzymes, such as amylases and proteases, are being added to animal feed to supplement the animals' own enzymes and improve digestion. Most animal feedstuffs are plant-derived and contain 'anti-nutritional factors' that interfere with digestion in a variety of ways. Many of these factors are non-starch polysaccharides, from cereal grains. Adding enzymes (beta-glucanases and arabinoxylanase) to feedstuffs increases the nutritional value of feeds by breaking down these factors, aiding digestion and making nutrients more easily absorbed.

Phytic acid is another anti-nutritional compound found in plant cereals that prevents dietary absorption of essential minerals such as iron and zinc. It also pollutes the environment because it is excreted as phosphorus in manure. Addition of phytase in the diet helps digestion, but until it was recently cloned this enzyme was not available in sufficient quantities for widespread use.

Many enzymes have been genetically modified to improve their useful properties. For example, subtilisin (the bacterial component that is the key enzyme ingredient in 'biological washing powders) has been modified to make it more stable at temperatures used for washing, and less likely to be inactivated by the calcium-chelating 'builders' that are required for good performance of detergents.

In manufacturing processes, enzymes often work at lower temperatures and pressures, and therefore save energy. Many enzymes are more efficient than their chemical counterparts, and do not require toxic chemical solvents to work, producing less waste and pollution.

Many enzymes are more specific, producing more of the desired product, and fewer by-products. This can be particularly important in the manufacture of drugs, where the side products may be toxic. Enzymes can carry out reactions for which chemical catalysts are unknown: these are finding uses for making specialty chemicals and new drugs.

Despite these advantages, which can make biological processes more environmentally friendly, much research and development is needed to make them cost-effective and able to compete economically with their chemical counterparts.

Genomics

Genomics is the study of a genome, rather than the study of an individual gene. For most of the last decade, genomics, and especially the Human Genome Project, have never been far from the headlines. However, even before the announcement in February 2001 that the sequencing of the human genome had been completed, the principles and technologies which had enabled this impressive achievement were being turned to the study of other areas of cell biology.

In biology, the suffix -omics generally refers to the study of a complete group or system of biomolecules. Just as genomics is the study of an organism's genome, proteomics is the study of an organism's entire complement of proteins. Phenomics is the name given to the science which attempts to integrate the information provided by all these areas of study into a holistic picture of the complete organism – its phenotype. As researchers focus on more and different groups of molecules, more 'omics will become part of the biological language.

The emergence of the 'omics has been made possible by advances in information technology and robotics. These technological improvements have allowed researchers to automate processes which previously had to be carried out by hand on the laboratory bench. Rather than studying one gene or protein at a time, this high-speed assembly-line approach permits huge amounts of data to be collected and stored in databases, to be analysed later. For example, the mapping of the human genome means we now have the DNA sequence of every human gene – however, we don't yet know what each gene does. It's analogous to having a map without all the place names on it. Now the race is on to uncover which proteins each gene codes for and what these proteins all do.

Proteomics

While the genome may be the blueprint for an organism, proteins are the structural and functional molecules required by virtually all life processes. Therefore, to truly understand how an organism functions we need to understand more than just its genome – we need to understand the proteome and all the other 'omes' as well. For this reason, proteomics is one of the fastest-growing areas of biological research now that the human genome has been mapped. However, unveiling the proteome is not as straightforward as it might appear.

Understanding what all the proteins in the body are and precisely what they do will give researchers a powerful new tool for diagnosis and treatment of disease. Many diseases are a result of defective genes, which create defective proteins that go on to cause problems for the organism. By being able to pinpoint the source of the disease, new treatments can be designed which precisely address the cause. For instance, treatments might be designed to specifically target the area of an abnormal protein which is causing dysfunction. Or, where insufficient production of an essential protein is the cause, artificially produced versions of the identified protein might be used to treat the illness – just as the protein insulin is used to treat diabetes.

Australian innovations in the field are already being utilised in pharmaceuticals and agricultural products – the anti-arthritis drug Remicade, for example, is based on a specific antibody patented by the Australian company Peptech. The 'omics also raise the prospect of personalised or targeted medicine – where the specific markers which distinguish an individual's disease can be identified and a treatment created to correct it. Because of the focused nature of this approach, adverse side effects can be reduced and treatment effectiveness improved.

Another aspect of personalised medicine is the potential to identify the specific gene variations a patient carries and using the knowledge to prescribe drugs known to be safe and effective for someone with their genotype. Currently, adverse drug responses occur in a significant proportion of patients, and determining the most effective pharmaceutical for an individual involves a lot of trial and error. This same technology could be used to predict a person's risk of suffering from diseases with a genetic component, such as heart disease or cancer, in a much more precise way than current predictive methods.

Despite all the potential benefits, there are also some serious social issues that will emerge as a result of this new technology. For example, the ability to accurately predict a person's individual risk of disease with an easy genetic test raises the prospect of health insurance companies insisting on such tests before issuing a policy – and even refusing to cover those who have a heightened genetic risk for say, heart disease or breast cancer.

Another problem is that personalised medicine and targeted drug treatments are currently very expensive. Many patients will be unable to afford such treatments on their own, and governments are already faced with difficult choices about which life-saving or life-improving drugs they can afford to subsidise. With many more such treatments on the horizon – and with a Pharmaceutical Benefits Scheme that is costing Australia nearly $5 billion dollars a year – the issue of which new treatments the government can afford to subsidise will only grow in importance.

Pharmaceutical Industry

Growth in biotechnology blurs the lines of differentiation between different fields of science. For instance, biotechnology advancements are integrating with semiconductor technologies such as micro-electromechanical systems resulting in the evolution of biochips. Biochips provide pharma companies with sophisticated tools for understanding gene expressions, biological mechanisms, speedy development of drugs, and accurate diagnosis. DNA and protein biochips streamline drug development and significantly lower average drug screening costs from $2 to $0.0001.

In addition, with the transition of the medical lab-on-chip (LOC) from a concept to a reality, the focus has been on developing high density and smaller-sized biochips wherein less than 300 wafers may accommodate the entire human genome. As opposed to conventional laboratory analysis, LOCs eliminate the need for trained personnel and expensive equipment to provide significant time and cost efficiencies.

The biotechnology industry by providing a comprehensive view of emerging technologies and applications such as bioanalytics, biopharmaceuticals, glycobiology, nanobiotechnology, and industrial biotechnology. Analysis of the technological trends, drivers, challenges and recent developments will assist in creating effective marketing and production strategies. Participants can identify potential collaborators, stay ahead of the competition, and stay shoulder-to-shoulder with critical developments in their industries.

The rapid evolution of glycomics as a natural extension of proteomics provides a better understanding of glycoproteins; thereby, helping the development of novel biodrugs. "Improved understanding of the structural and functional data in glycomics coupled with phenomenal developments in genetic engineering techniques drives the usage of recombinant glycoproteins as therapeutic agents and as an alternative to small molecules," says the analyst of the study.

Demand for therapeutic glycoproteins and monoclonal antibodies is expected to increase creating an urgent need for higher production capacity. Additionally, alternate manufacturing mediums such as transgenic animals, plants, and mammalian and fungal cell lines are also on the rise. Fungal cell lines, in fact, provide considerable time and cost benefits over mammalian cell lines. The latter is not only a lengthy process but may also alter the properties of the final therapeutic glycoprotein. Conversely, fungal cell lines such as engineered yeast expression systems facilitate production of humanly glycosylated protein, provide faster fermentation and generate higher product yield.

Fermentation Process

The term "FERMENTATION" has numerous contextual meanings, but in relationship to ethanol production it refers to the metabolism of carbohydrates under anaerobic conditions. Any simple or complex carbohydrate can be fermented to produce ethanol. Usually this fermentation is carried out by certain ubiquitously distributed yeast, but a few bacteria are also able to produce ethanol in commercial quantities.

The commercial or industrial production of ethanol is produced as follows:

- A grain, usually barley, which is rich in complex glucose polymers (complex carbohydrates) is collected and wetted.
- As it is stored in a warm, dark place the seeds germinate (sprout) and release enzymes that break down or hydrolyze the #polysaccharides to simple sugars which the yeast can metabolize. This process is called MALTING.
- The malt is dried and crushed to improve the extraction of the sugars. The dried material can be stored at this stage. This is what you can buy in a store if you want to brew your own booze.

— In the next step, called MASHING, the sprouted malt is suspended in water where the enzymes continue to break down the polysaccharides to release simple sugars.
— Following mashing, the liquid, or MALT WORT or WORT, containing the dissolved simple sugars is separated from the insoluble material.
— Hops (a plant grown in the US mainly in Washington) are added to the wort, which is boiled to destroy extraneous enzymes, to extract the hop-flavors and to precipitate the proteins which could add unwanted flavors and cloudiness to the final product. The hop-flowers contain substances that *inhibit spoilage microbes* and aid in the final clarification of the product.
— Yeast are then added (PITCHED) and the mixture is placed in large closed containers, usually metal (copper or stainless steel tanks today), in large commercial operations, so the fermentation can proceed in the absence of air.
— The fermentation goes on for about 7 days at a cool temperature which is optimum for ethanol production.
— The brew (green beer) is may be aged (LAGERED) under a variety of conditions for various times as desired by individual brewers. During the aging, chemical changes occur spontaneously that subtly alter the flavor of the product. However, most brewers consider that fresh beer is the better product as the flavor changes in beer as it reacts with even small quantities of oxygen are considered damaging to the flavor.
— Finally the product is usually cleared (the yeast removed), bottled and pasteurized.

Non-Food Industrial Microbiology

Microbes have been used for about 100 years to produce

industrial chemicals for human use. The microbes are cultivated under rigorously controlled environmental conditions conducive to optimum production of the given product in rather humongous fermenters. Fermenters are tanks that may hold 1,000 of gallons, or more, of a culture. They must be made of materials, usually stainless steel, that can be heat sterilised and which will not react with the microbes or with the desired products. They must be able to be tightly sealed to prevent contamination and yet must contain numerous openings for monitoring the progress of the fermentation and for controlling the internal environment.

All industrial microbial processes deal with similar problems:

— Finding the *least expensive medium* in which to grow the microbe so as to maximise yield and profits.

— Often this is a waste product from another industrial process, such as corn steep liquor, sugar processing wastes or whey.

— Maintaining strain purity and developing better strains for improving the yield.

— A single mutation may decrease the yield by a significant percentage or result in undesirable substances being produced. The industrial research laboratories constantly seek better strains for the production of their product.

— Preventing contamination by other microbes and by viruses (phage) that live on the microbe involved.

— The media must be sterilised prior to being inoculated with the desired organism and purity must be maintained throughout the production process. A small quantity of a contaminant may produce an enzyme that can destroy the product in 1,000s of gallons of medium. For many microbes, viruses present a constant danger as a single virus can infect and destroy the desired microbe in an entire tank. The sterilisation of large containers and huge quantities of media represent both an engineering and microbial challenge.

— Developing rapid and efficient methods for purification of the desired produce in a stable form that is safe to use.

— Always striving to improve yield by modifying the strain, nutrients or environmental conditions.

— As product yields are exquisitely sensitive to subtle modifications in the nutrient and the environmental conditions, these are constantly monitored For example, the pH, oxygen content, nitrogen/phosphorous ratio etc. may be adjusted during the production process.

— Safe and inexpensive disposal of the massive quantities of waste products remaining after the product is formed.

— The waste products of these large fermentations present major waste disposal problems as they are rich in organic matter that are highly polluting if released untreated into the environment. However, the cost of treatment cuts into the profit margin and increases the cost of the product.

REFERENCES

Cooney, J.J., and Gadd, G.M., "Metals and microorganisms," *Journal of Industrial Microbiology* 14(2, 3, 4), 59 353, 1995.

Sayler, G.S., Fox, R., and Blackburn, J.W., "Environmental Biotechnology for Waste Treatment," Plenum Press, New York, 1991.

Shannon, M.J.R., and Unterman, R., "Evaluating bioremediation," *Annual Review of Microbiology* 47, 715 730, 1993

Bibliography

Ajello, L., L.K. Georg, W. Kaplan, and L.Kaufman, "Laboratory Manual for Medical Mycology", *U.S. Publ. Health Serv. Publ.*, 1963.

Angelotti, R. and MJ Foter, "A Direct Surface Agar Plate Laboratory Method for Quantitatively Detecting Bacterial Contamination On Nonporous Surfaces", *Amer J Public Health.* Sept:170-174, 1957.

Beneke, E.S., "Medical Mycology Laboratory Manual", Burgess Publ. Co.,Minneapolis, 1958.

Beuchat, L.R. and A.D. Hocking, "Some considerations when analysing foods for the presence of xerophilic fungi", *J. Food Prot.*, 1990.

Booth, C., "Fusarium", *Laboratory Guide to the Identification of the Major Species,* 1977.

Brack, A. (Ed.), *The Molecular Origins of Life: Assembling Pieces of the Puzzle, Cambridge University Press, London,* 1998

Brock, T. D., *Milestones in Microbiology*, Prentice-Hall, Englewood Cliffs, NJ, 1961.

Burgess, L. W., C. M. Liddell and B. A. Summerell, *Laboratory manual for Fusarium research*, 2nd ed. University of Sydney, Sydney, 1988.

Burleson F.G., T.M. Chambers and D.L. Wiedbrauk, *Virology: a laboratory manual*, Academic Press San Diego, Sydney, 1992.

Casadesús, J. & R. D'Ari, "Memory in bacteria and phage", *BioEssays*, 2002.

Cooney, J.J., and Gadd, G.M., "Metals and microorganisms," *Journal of Industrial Microbiology* 14(2, 3, 4), 59 353, 1995.

Dreyfus, D. H., "Evidence suggesting an evolutionary relationship between transposable elements and immune system recombination sequences", *Molecular Immunology*, 1992.

Dring, D.M., "Techniques for microscopic preparation", *In* C. Booth (ed.), *Methods in Microbiology*, vol. 4. Academic Press, New York, 1971.

Farries, T. C., Steuer, K. L. and Atkinson, J. P., "Evolutionary implications of a new bypass activation pathway of the complement system", *Immunology Today*, 1990.

Gessner, M.O. and Newell, S.Y., "Biomass, growth rate, and production of filamentous fungi in plant litter", *Pages 390-408 in: Manual of Environmental Microbiology*, Second Edition, ASM Press, Washington, D.C., 2002.

GreA and DksA. Potrykus, K., D. Vinella, H. Murphy, A. Szalewska-Palasz, R. D'Ari & M. Cashel, "Antagonistic regulation of *E. coli* ribosomal RNA *rrnB* P1 promoter activity" Journal of Biological Chemistry, 2006.

Haldane, J.B.S., "Sampling errors in the determination of bacterial or virus density by the dilution method". *J. Hygiene.*, 1939.

Hanson, J.R., J. L. Macalady, D. Harris and K.M. Scow, "Linking toluene degradation with specific microbial populations in soil", *Applied and Environmental Microbiology*, 1999.

Hauschild, A.H.W. and R. Hilsheimer, "Enumeration of food-borne Clostridium perfringens in egg yolk-free tryptose-sulfite-cycloserine agar", *Appl. Microbiol.*, 1974.

Hendolin P, Markkanen A, Ylikoski J, Wahlfors J.J., "Use of multiplex PCR for simultaneous detection of four bacterial species in middle ear effusions", *Journal of Clinical Microbiology*, 1997.

Hugo W. B., and A. D. Russell, *Pharmaceutical microbiology*, 3rd ed Singapore: P.G. Pub., 1983

Huws, S. A., Smith, A. W. Enright, M. C., Wood, P. J. and Brown, M. R. W., "Amoebae promote persistence of epidemic strains of MRSA", *Environmental Microbiology*, 2006.

Jetten et al, "Microbiology and application of the anaerobic ammonium oxidation (anammox) process", *Environmental Biotechnology*, 2001.

Marchalonis, J. J., Jensen, I. and Schluter, S. F., "Structural, antigenic and evolutionary analyses of immunoglobulins and T cell receptors." *Journal of Molecular Recognition*, 2002

McGinnis, M.R., *Laboratory Handbook of Medical Mycology*, Academic Press, New York, 1980.

Pelczar, Michael J., *Microbiology*, New York: McGraw-Hill, 1958.

Sayler, G.S., Fox, R., and Blackburn, J.W., "Environmental Biotechnology for Waste Treatment," Plenum Press, New York, 1991.

Shannon, M.J.R., and Unterman, R., "Evaluating bioremediation," *Annual Review of Microbiology* 47, 715 730, 1993

Smith, A. W., "Biofilms and Antibiotic Therapy: Is there a Role for Combating Bacterial Resistance by the Use of Novel Drug Delivery Systems?", *Advanced Drug Delivery Reviews*, 2005.

Vinella, D., C. Albrecht, M. Cashel & D'Ari, R., "Iron limitation induces SpoT-dependent accumulation of ppGpp in *Escherichia coli*", Molecular Microbiology, 2005.

Index